Management of Data in
Clinical Trials

Management of Data in Clinical Trials

ELEANOR McFADDEN, M.A.

Eastern Cooperative Oncology Group
Frontier Science and Technology Research Foundation

A Wiley-Interscience Publication
JOHN WILEY & SONS, INC.
New York • Chichester • Weinheim • Brisbane • Singapore • Toronto

This text is printed on acid-free paper. ⊛

Copyright © 1998 by John Wiley & Sons, Inc. All rights reserved.

Published simultaneously in Canada.

Library of Congress Cataloging in Publication Data:

McFadden, Eleanor, 1948–
    Management of data in clinical trials / Eleanor McFadden.
      p. cm. — (Wiley series in probability and statistics.
  Probability and statistics)
    Includes bibliographical references and index.
    ISBN 0-471-30316-X (alk. paper)
    1. Clinical trials—Data processing. I. Title. II. Series.
      [DNLM: 1. Clinical Trials—methods. 2. Database Management
  Systems. 3. Research Design. QV 771 M478m 1997]
  R853.C55M39 1997
  610′.72—dc21
    DNLM/DLC   97-1251
  for Library of Congress

                                CIP

Printed in the United States of America
10 9 8 7 6 5 4 3 2 1

For my mother and in memory of my father

Contents

Preface

In this text I have tried to provide a general overview of the steps involved in managing data in clinical trials. The information should be of use to anyone who is working in the field of clinical trials, but particularly those who are working with trial data. This includes Clinical Research Associates, Data Coordinators, physicians, nurses, and statisticians. In my experience, most of these individuals receive little training in the practical aspects of clinical trials, and while sound in theory, they are often at a loss when it comes to details. I have found a lack of published material covering this field, and I have written this book to fill the existing gap.

As well as discussing the more traditional aspects of data management—the design and completion of case report forms, I have included information on the planning phase of a trial, use of computers and other technology, training and education, and the implementation of Good Clinical Practice. Much of what I have included is based on my own experience in the field of data management, and the questions that I have frequently been asked. For the most part the chapters follow the life of a trial from the design stage to the analysis stage, with emphasis on the systems that are needed for managing data. While my own experience has been primarily with cancer clinical trials conducted in the United States, I have tried to make the information general and applicable to all kinds of trials. If I was aware of differences in systems for different types of trials and trials done in different countries, I have pointed out these differences in the text.

The goal of the book is to help you to manage trial data in a way that ensures the timeliness and integrity of the data collected. Not every chapter will be relevant to everyone who reads it, but my hope is that all readers will find some information in the book that will assist them in their clinical trials environment.

Acknowledgment

In 1992 I was invited to work with a group of individuals in the preparation of a series of manuscripts on data management for a special edition of Controlled Clinical Trials. The edition was finally published in 1995. That collaboration expanded my knowledge of clinical trials beyond my own specialized area of cancer trials, and showed me the similarities and differences between cancer trials and other disease areas. The idea of this text originated during the collaboration, and many of my suggestions in this book are enhanced by the final publications and the knowledge freely shared by my colleagues in that project, in particular, my primary coauthor, Fran LoPresti.

My knowledge of clinical trials in countries outside the United States was enhanced when I participated in two workshops organized by the European School of Oncology and the European Organization for Research and Treatment of Cancer. Di Riley co-chaired both these sessions with me, and I benefited from her experience with clinical trials in the United Kingdom. I would also like to thank the reviewers appointed by John Wiley & Sons for their valuable comments. I hope that I have addressed them adequately in the final version of the book.

There are three individuals to whom I owe a special debt. During my 20 years in the Eastern Cooperative Oncology Group (ECOG), I have had the privilege of working with two statisticians who have themselves made many important contributions to the design and conduct of clinical trials—Marvin Zelen Ph.D. and David Harrington Ph.D. The third person, Paul Carbone M.D., served as the Group Chair of ECOG for 20 years and is truly a pioneer in developing new treatments for patients with cancer. Without the support and guidance of these three individuals, I would not have gained the knowledge that has led to this book. To them and all the colleagues that I have worked with at ECOG, I extend a heartfelt thanks.

CHAPTER 1

Introduction

Clinical trials are utilized in many clinical specialties to test the efficacy of a specific treatment or intervention in a group of patients, and inferences are then drawn about the use of the treatment in the general population. There are different types or phases of clinical trials, but they all have one thing in common. The results of the completed study are only as good as the data collected and analyzed during the trial. A "good" result of a clinical trial is one that provides the *correct* answer to the questions asked, and this answer may not necessarily be one that is positive or statistically significant. Good data management practices are essential to any clinical trial, yet this is an area often neglected or glossed over when planning a trial. This book discusses the various stages of a clinical trial from planning to analysis and closeout, and provides guidelines for the management of data to answer the trial objectives.

Clinical trials can be large or small; they can involve one clinical center or multiple clinical centers. Multicenter trials allow more rapid accrual of patients to a trial and therefore a faster answer to the question being asked. The results of multicenter trials are also more easily generalized to the population as a whole because the results are obtained from trials conducted in a variety of settings. Large multicenter trials usually have a Coordinating Center with a wide range of responsibilities, including input in trial design, quality control and computerization of data, interim and final analyses of the data, and preparation of a report on the results. The Coordinating Center may also develop systems to ensure the smooth flow of information among the people involved in the trial. In trials done at single centers, these functions are often the responsibility of the investigator who designs the study. Regardless of the size or complexity of a trial, detailed planning is essential, and the guidelines in this book have relevance to even the smallest clinical trial.

There is no one "correct" way to conduct a clinical trial. There are many different ways to organize a trial, and choices need to be made based on the environment and resources available. The system developed for conducting a trial should be based on intelligent decisions after reviewing the study requirements in great detail. Careful prospective planning is essential to ensure that the study runs smoothly, that all necessary data are collected in a timely way, that on-going progress can be monitored to ensure patient safety, and that final results can be analyzed and published as soon as possible after the termination of the study. While everyone involved in clinical trials may think that their way of doing things is the best way, in reality a data management system is successful if, using available resources, it results in the collection of complete, timely, accurate data that answer the scientific questions.

DEFINITION OF A CLINICAL TRIAL

Throughout this book, a clinical trial is defined as a trial involving the assessment of one or more regimens used in treating or preventing a specific illness or disease. The regimen may be curative, palliative, or preventive. There are other types of clinical studies, some involving the administration of questionnaires, surveys, or specific tests to subjects who fulfill certain requirements. These studies collect information on the subjects entered but do not assess the effects of interventions. Many of the guidelines for therapeutic trials apply equally to these kinds of studies. For the most part, in this book examples and terminology will refer to therapeutic trials, but parallels may be drawn for other types of studies.

TYPES OF CLINICAL TRIALS

The design of a clinical trial depends on the objectives and the experimental treatment. Some trials involve comparisons with other treatment regimens, and other trials seek to further knowledge about the effects and effectiveness of a specific treatment. There are four traditional types of therapeutic clinical trials:

Phase I
Phase I trials are small noncomparative studies that test new therapies in humans, usually without therapeutic benefit to the patient. The objective of a Phase I study is to find the optimal dose or maximum tolerated dose (MTD) to use in further testing of the treatment. The MTD is defined as that dose which can be administered without inducing unacceptable side effects. Rapid reporting and assessment of all toxicities is therefore critical in any Phase I study. The study design will require that a specific number of patients be

entered at a dose level. If side effects are acceptable for those patients, then a higher dose will be tested. If accrual is rapid, the trial should be suspended pending evaluation of side effects at one dose level before treating patients with the higher dose, and additional patients should not be treated at the same or a higher dose until this evaluation is complete.

Phase II

Phase II trials are noncomparative trials that assess the therapeutic activity of new treatments in humans. The objectives are to identify promising new treatments that can then be moved into the next phase of testing in a larger population. As with Phase I studies, timely reporting of outcome data is critical. This data will include assessments of treatment efficacy and treatment-related toxicity or side effects. Phase II trials can be randomized if two or more new treatments are available for testing in the same patient population, but analysis usually will not involve comparisons between the arms. Each arm is assessed independently for therapeutic activity according to the criteria defined in the protocol. Phase II trials are usually small and often have a two-stage design, where a pre-set number of patients is entered and assessed for positive outcome. If enough patients satisfy the criteria specified in the study design to determine that the treatment is effective, additional patients are entered to complete the total accrual goal of the trial.

Phase III

Phase III trials are large studies with more than one treatment arm. They are comparative trials, that is, among the treatment arms. Most Phase III trials involve a random assignment between a control arm and one or more experimental treatment arms. The control arm usually is the standard care for a disease, and can be observation and monitoring without administration of any therapy. In some Phase III trials, the patient (and often the physician) are "blinded" to the treatment assignment and do not know what treatment the patient is receiving. Usually the control arm for these trials involves a placebo which is identical in appearance to the medication that is being tested as the experimental treatment. Trials involving a placebo are feasible only when the experimental treatment arm does not cause severe or unusual side effects. In Phase III trials the randomization to one of the available treatments is done prospectively. There will be a mechanism in place for patients to be registered before starting treatment, and the treatment is assigned randomly using an algorithm defined by the study statistician. It is important to note that the treatment assignment is not under the control of the treating physician. Randomization raises practical and ethical issues which are discussed in more detail in Chapter 6. The benefits of randomization versus the use of historical data for comparison of treatment effects is a subject of debate in the statistical literature and is not covered here. In this book, all discussions about Phase III trials refer to prospective, randomized studies.

Phase IV
Phase IV trials are postmarketing surveillance trials for collecting additional information on short- and long-term side effects of treatments.

DEVELOPMENT AND CONDUCT OF A CLINICAL TRIAL

A clinical trial goes through various stages from the development of the hypothesis to be tested to the analysis of the results. In very broad terms, the three stages of a clinical trial are:

- Design and development
- Patient accrual and data collection
- Follow-up and analysis

In each of the three stages, consideration must be given to techniques for management of data. These three stages will be covered extensively in subsequent chapters, but a general outline is useful here.

Design and Development

During the design and development stage of a trial, a protocol document is developed. The protocol contains critical information for the participants in the trial with sections describing the scientific rationale for the trial, defining the patient population, details of the treatment plan, criteria for assessment of effectiveness of the treatment(s), and other administrative and scientific information that is important to the conduct of the trial. This document becomes the rule book for the trial and ensures that the defined patient population will be treated in a uniform way. It is also important to develop mechanisms to enforce and monitor compliance with the protocol once the study is open to patient accrual.

In parallel with the development of the protocol, the data to be collected to answer the study objectives must be defined. Decisions must be made about how the data are to be collected, whether on paper or electronically. Whichever is used, a format for the data capture forms or screens (or both) needs to be designed. In using paper forms, a system must be set up for distributing the blank forms to participants and for ensuring that completed forms are returned in a timely way to the Coordinating Center or the individual responsible for quality control. If electronic data submission is to be used, hardware and software must be in place and tested at all participating sites. This may involve development of specific study software. If collected samples (e.g., blood, X rays) are part of the study, mechanisms for the collection and receipt of samples, and for recording of results of central review, need to be designed. If these materials are to be sent to Reference Centers for review, then communication systems need to be defined between the Coordinating and Reference centers so that there can be rapid transmission of results. If the Reference Center also must communicate directly with the participating sites, that system needs to be established.

An important part of a trial is the system used for patient entry, and this also needs to be planned in advance. There are often requirements to collect data that document compliance with regulations or other administrative data needed during the trial. This can be as simple as building a roster of names, addresses, telephone, and fax numbers of all participants, but such information should be identified in the development phase of the trial so that the conduct of the study is not compromised once under way. Decisions need to be made about the use of computers and other technology, and all relevant systems need to be in place and thoroughly tested.

During this stage, documentation of standard procedures is developed. Among such documentation are the policies and procedures within the Coordinating Center and also for participating sites. It is also useful to develop time lines for the entire trial to ensure that the trial stays on schedule. This can include time lines for development of the protocol, development and testing of forms, patient recruitment, data collection, and resolution of problems as well as for analysis. While certainly adjustments will be made as the trial proceeds, a preliminary plan helps identify problem areas more quickly.

There are many things to consider at this design and development stage, and it is indeed the most critical stage of a clinical trial. This stage is important for all clinical trials regardless of size or complexity. Activating a study without proper systems in place can lead to inadequate and incomplete data and can compromise the integrity of the trial.

Patient Accrual and Data Collection

After a trial has been activated, emphasis switches to patient registration, data collection, and quality control, and the mechanisms for these activities should have been set up during the design and development phase. There needs to be close monitoring of the trial to ensure that accrual rates are acceptable, that the eligibility requirements for the study are realistic, that regulatory requirements are being met, and that there are not unexpectedly high rates of adverse reactions to treatment. Ongoing quality control of all data collected is done to check for consistency and completeness, and there should be a system for ensuring that data are collected and submitted in a timely way. As well as routine ongoing monitoring of the study, interim statistical analyses should be done according to the design specified in the protocol. If in addition the Phase III trials are monitored by a Data Safety and Monitoring Committee, the ongoing reports need to be prepared for that committee so that they can fulfill their responsibilities on reviewing safety and, when appropriate, treatment effects.

Follow-Up and Analysis

Once the accrual goal has been reached and the trial is closed to further patient entry, it enters the follow-up phase. The length of this phase depends on the study design and study end points. Adequate time should be allowed for complete data to be submitted and for the data to mature sufficiently for the results to be meaningful. Disclosure of premature results can lead to false conclusions about study results. During this stage of the study, it is important that the data are known to be complete and accurate and all queries resolved. Any clinical review should be completed before the final analysis. Many trials also require closeout procedures that involve archiving of electronic and paper files.

COORDINATING CENTER

As mentioned previously, trials can involve multiple centers with a central Coordinating Center. Figure 1.1 shows the usual way that data flow is repre-

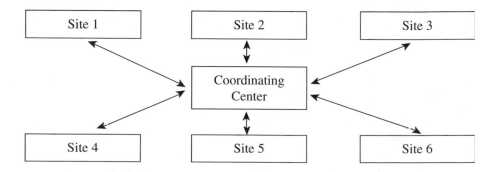

Figure 1.1. Flow of Information through a Coordinating Center.

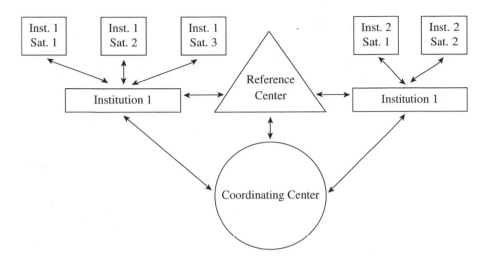

Figure 1.2. Multicenter Trial with Satellite Institutions and a Reference Center.

sented for such studies. All data flows from the participating sites directly to the Coordinating Center. Figure 1.2 gives a more complex model where satellite sites send data to a main site from where it is forwarded to the Coordinating Center, and where the sites send data to a Reference Center, such as an X-ray reading facility. That facility then forwards results to the Coordinating Center. It is essential for all models to ensure that data flow smoothly and can be tracked through the system.

The Coordinating Center is responsible for the overall conduct of the trial, for ensuring that the trial is proceeding as planned, that all necessary systems and mechanisms are in place and functioning, that the protocol is being followed and necessary data are being submitted, and that all regulations are being met. Many of the key personnel for the trial are located at the Coordinating Center, and they are involved in all phases of the trial from development to analysis. The Coordinating Center is the main center for communications between all participants, so it must ensure that participants receive notification of changes to the protocol, changes to forms, patient safety updates, and other critical information. The Coordinating Center must also track data and materials collected for the trial. The systems become more complex when materials have to be sent to Reference Centers for review or processing. If this is required, decisions are made about whether the materials should be sent directly from the site to the Reference Center or sent to the Coordinating Center and then forwarded from there. This may depend on the urgency of the review or the type of materials being collected. Direct transmission to the Reference Center is more efficient but makes it more difficult for the Coordinating Center to keep track of the materials. Quality control systems need to be developed and implemented at Reference Centers to ensure consistent and objective review of the materials.

In this book, the descriptions and procedures are described using the Coordinating Center model, although it is recognized that small trials can be done in a single institution, with all functions of the Coordinating Center being carried out at that site by one or more people. The Coordinating Center is defined as the place where data forms are collected, quality control is done, and data are computerized and analyzed. The participating site or institution is defined as the place where patients are screened and entered on the trials, and where source data are collected and transcribed on to data collection forms.

Usually a large clinical trial (and smaller ones too) is sponsored and funded by a pharmaceutical company or a government agency. Besides providing financial support for the trial, the Sponsor may provide study drugs or other materials. The Sponsor may use its own personnel and facilities for the Coordinating Center or may contract this function to another organization. If contracted, the Coordinating Center acts as the agent of the Sponsor in the conduct of the trial, although it is the responsibility of the Sponsor to ensure that all of its requirements are being met.

Personnel

There are many people involved in the conduct of a large clinical trial, and all the relevant people should be involved at all stages, starting with the development phase. In different organizations, different titles are used for these people. For clarity, the following definitions apply in this text:

Study Chair
The Study Chair develops the scientific concept to be tested and is usually a clinician. This person is responsible for the ongoing conduct of the trial and often will review some or all of the submitted data. Other terms used for the person supervising the overall trial plan are Study Coordinator, Study Principal Investigator, and Clinical Coordinator. In some clinical trial organizations, this role is filled by a Trial Team.

Statistician
The statistician is involved in the design of the study and is responsible for calculating the sample size and defining the statistical methodology that will be used in the trial and analysis. Throughout the study the statistician is responsible for analysis of the trial data and is involved in monitoring the progress of the trial.

Clinical Research Associate

The Clinical Research Associate (CRA) is the person at the participating site who is responsible for completing the study case report forms (CRFs) and submitting them to the Coordinating Center. The CRA's responsibilities usually are more extensive than completing the required data forms and can involve patient entry, scheduling of visits and tests, and preparing required regulatory documents. Other terms for this person include Data Manager or Data Coordinator. Clinical Research Associates should be involved to some degree in the design and pilot testing of case report forms and in the evaluation of proposed systems and procedures. If software applications are to be used at the participating sites, Clinical Research Associates should play a part in thoroughly testing the applications prior to the activation of the trial.

Data Coordinator

The Data Coordinator is responsible for quality control of data in the Coordinating Center. This person is also responsible for generating edit queries and data requests, for processing patient registrations, and for maintaining all study files. The Data Coordinator assists the Statistician in preparing data sets for analysis and is the primary contact with the trials personnel at the participating sites. For a small, single-center trial where there is no Coordinating Center, the person fulfilling this function may be the Clinical Research Associate, but because it is important to distinguish between the two roles, both are used in this book. Other titles used are Data Manager and Data Specialist. The Data Coordinator should be involved in the design of case report forms, review of the protocol document, and development of the systems and procedures.

Database Administrator

The Database Administrator (DBA) is responsible for designing and setting up the computer database and for ongoing maintenance, including installation of software upgrades. The Database Administrator also ensures database's integrity and security and is responsible for maintaining an adequate backup system.

Systems Analyst

The Systems Analyst is responsible for the design of the trials software and for overseeing development and testing.

Programmer

The Programmer is responsible for writing and maintaining the programs under the direction of the systems analyst. In a clinical trials environment there may be programmers for database programming and others for development of statistical programs.

With the exception of the computing staff who are only involved if a study is computerized, all these individuals need to be involved in the trial from the start. For large trials these people will all have separate responsibilities, but, for small trials, one person may handle all or several of these functions. This book focuses on the responsibilities of the Clinical Research Associate and Data Coordinator in the design, conduct, and analysis stages of a clinical trial.

TRAINING AND EDUCATION

It is important, particularly for large Phase III trials, to establish a mechanism for initial training for participants and also for ongoing education. Training can be done by having participants attend a trials workshop, by video or with written materials. Whatever mechanism is used, emphasis should be placed on ensuring that the participants understand the protocol and the requirements of the trial. Once the trial is underway, continuing education can be achieved by any of the above methods. A periodic newsletter to participants can maintain interest in the trial and ensure that important information is conveyed.

REGULATIONS AND ETHICS

Most countries have regulations that govern the conduct of clinical research within that country, and it is essential that participants be aware of regulatory and ethical requirements before embarking on the trial. The rights of the patient in the trial must be adequately protected, and it is important that these rights are understood. They include the patient's right to withdraw from the trial at any time and also protect the patient's confidentiality. If regulatory documents have to be filed before a trial starts, the Coordinating Center should be responsible for developing a mechanism to ensure that this happens. Additional information is provided in Chapter 9, Data Management and Good Clinical Practice.

SUMMARY

To be successful, it is important that a clinical trial be well designed and planned before entry of patients. The planning stages of the trial are critical, and many things need to be considered. While these are more critical for a large multicenter trial, they are also important for a small trial being done in only one institution. There are many things that can go wrong during a trial, but with adequate planning and the appropriate resources available, many can be avoided. Many trials have been successfully completed, and there are many organizations involved in the conduct of controlled clinical trials. Before embarking on a large trial, there is much to be learned from the systems already in place.

CHAPTER 2

Study Design and Planning

This chapter identifies some of the decisions to be made when planning a trial and provides a summary of the key areas to consider. Careful planning is essential for all trials, large or small, simple or complex. Leaving critical decisions until after the study is activated or, even worse, until after patient enrollment is complete, can seriously jeopardize the outcome of a trial. This may seem like an obvious statement, but investigators are often anxious to activate a trial as quickly as possible, and pressure can be put on the trial team to begin patient accrual before everything is ready. Such pressure should be resisted, and studies should only be activated after all necessary systems are in place. The planning phase should include discussion of the full scope of the project and the support mechanisms that will be needed during the conduct of the trial. This includes identification of those who will be involved and at what time points, and the definition of their responsibilities. It also includes planning the flow of information between participants, indicating how data will flow and how feedback will be provided to sites that enter patients in the trial. Decisions made during this stage of the trial have a major impact on the quality and completeness of the data which will ultimately be analyzed to determine the results of the trial.

STUDY DESIGN

The first step in any clinical trial is to identify the scientific question being asked and to state the objectives of the trial. This is usually done by the Study Chair, either alone or as a member of a trial team. Once the objectives have been defined, it is important to assess whether the trial is feasible. The Statistician on the study team should work with the Clinician to prepare the statistical design for the study and calculate the sample size needed to answer the question(s). Then there should be assessment of whether sufficient patient resources are available to ensure accrual in a reasonable period of time. A poll can be taken asking prospective participants about likely accrual rates from their site, but investigators tend to overestimate the numbers of patients that they think they will be able to enter. If possible, it is therefore useful to review previous trials in the same population and to assess accrual rates from these historical data. If there are an insufficient number of patients available at a clinical site, it may appropriate to ask other sites to participate in a multicenter trial. Once the feasibility of doing the trial has been established, a protocol should be developed.

PROTOCOL DEVELOPMENT

A protocol is a document that describes a clinical trial in detail and provides information and rules for the conduct of the trial to those involved. A protocol should be complete, clear, and consistent and made available to all participants prior to the activation of the trial. The protocol should contain sufficient details about the trial to ensure that there is uniformity in the selection and treatment of patients entered on that trial. If the same participants will be involved in multiple trials over time, there are advantages to maintaining consistency in the format and contents of protocol documents, since it allows people to become familiar with the documents and to be able easily to locate information in them. The protocol should prospectively address the entire conduct of the study, both scientific and administrative. The following sections are recommended for clinical protocols.

Study Objectives

This section should clearly define the questions to be addressed in the proposed study. There may be primary objectives and secondary objectives, and it is important to be realistic in defining these objectives and to limit the questions being asked to those that can be answered in a reasonable period of time with the available patient population. For example, if there are approximately 50 new patients per year with the relevant characteristics for the trial, it would not be practical to define multiple scientific objectives that require 1000 patients to be entered before there are sufficient data to answer the questions being asked. This trial would accrue patients for more than 20 years, by which time the questions being asked would probably be irrelevant.

Background

This section introduces the concepts behind the trial. It should contain information and references about prior research and observations that led to the proposed study. Since protocol documents are often reviewed by many people, it is advantageous for this section to summarize as much of the relevant background information as possible while keeping to a reasonable length. This is particularly important in a multicenter trial where people not directly involved in the development of the protocol may decide whether or not to participate in the trial based on the information given in this section.

Eligibility Criteria

This section defines the patient population to be studied in the trial. It should describe the population well enough so that eventual results can be reasonably interpreted. Problems can occur when the patients entered on a trial have so many variable characteristics that no meaningful conclusion can be drawn when the data are analyzed. Conversely, the criteria should not be so restrictive that it is almost impossible to find eligible patients.

The eligibility criteria should be clinically relevant to the protocol treatment. For example, if the trial involves a drug known to cause cardiac side effects, patients with a history of cardiac disease should probably be excluded from entry. The eligibility section should be prepared primarily by the Study Chair, and each criterion should be justified. The criteria listed for eligibility should take into account the timing and cost of required tests as well as the feasibility of testing the criteria at all participating sites. For example, if a CT scan has to be negative before a patient can be entered, it is important to be sure that all participating sites have equipment available to do the CT scans. It may also be necessary to determine whether the cost of tests will be covered by medical insurance or whether the patient will have to bear the cost. If certain tests have to be done within a specified period of time prior to entry on the trial, this information should be stated in this section.

Eligibility criteria should describe the characteristics of the patient at the time of entry on to the study and should, as far as possible, require objective or quantitative answers. Events that occur after entry on to the trial should not affect eligibility assessment. For example, overdue or missing case report forms should not make the patient ineligible nor should changes in clinical characteristics after registration. Such changes may make a patient unevaluable for the objectives of the study or, if serious, could mean that protocol treatment should not be given, but if they occur after registration, they do not affect the determination of eligibility. Eligibility criteria that require subjective assessment such as life expectancy should be avoided as much as possible.

In defining the patient population, the criteria should be clear and unambiguous. While the eligibility section is normally written as a list of requirements for a patient to be eligible for the study, it is sometimes beneficial to also include a list of criteria that make a patient ineligible. If both options are used, the two sections should be treated as mutually exclusive and not as one merely being the negative of the other. Sometimes there are requirements that sound awkward if worded as criteria for patient eligibility, and these can be written more clearly if they are expressed as a criteria that would make a patient ineligible.

Registration Procedures

This section describes how to enter patients on to the trial. Chapter 6 gives more information on possible systems for patient registration, but regardless of the system used, it is strongly recommended that only prospective registrations be allowed (i.e., patients must be officially entered on the trial *before* treatment is begun).The registration instructions for the protocol should tell participants what procedures to follow to enter a patient on the study. Included could be a telephone number and hours of operation if registrations are done over the phone, or software instructions if a computer is used. As well as telling the participants what steps to follow, it is useful to list any information which the participant needs to have ready to complete the registration. For example, if all eligibility criteria are to be confirmed prior to entry, asking the participant to complete an eligibility checklist prior to registration should ensure that all necessary information will be available at the time of the call. A Registration Worksheet could be developed for the institutions listing all the things that need to be done prior to initiating the registration. An example of a Worksheet is given in Figure 2.1.

Trial 0101   REGISTRATION WORKSHEET
Patient Initials: _____

Prior to calling to register a patient on this trial, please be sure that the following are done/available:

1. Patient has signed a consent for the trial.
2. Eligibility Checklist is completed and signed by MD.
3. Ethics Committee Approval is current.
4. Pharmacy has adequate supply of drug.

The following data will be needed during the call:
Date of Birth: _____
Hospital ID Number: _____

Figure 2.1. Registration Worksheet.

If the trial requires multiple registrations/randomizations at different time points during the call, procedures and information for all these registration steps should also be included in the protocol document.

Treatment Administration

This section of the protocol defines the treatment plan for the trial. For therapeutic trials, this includes information about the treatment regimen(s), whether they involve drugs, surgery, or other therapeutic modalities. This section should be as complete and clear as possible, since it is important that all patients entered on the trial follow the same treatment plan. This will allow valid interpretation of the data when the analysis is done.

Besides a detailed plan for treatment, this section should also give instructions to be followed if the patient experiences side effects from the protocol treatment. Should modifications be made to the regimen, or should further treatment be delayed until the side effects disappear? Should supportive medications be given to counteract the side effects? If the side effects continue even after a dose reduction, then should the dose be reduced again? If so, it is important to clearly specify how the reduction should be made. Often these reductions are expressed as a percentage of the original dose; should subsequent percentage reductions (or increases) in dose be recommended as part of the protocol treatment plan, it is important to specify whether the second percentage should be based on the original dose or on the dose after the first reduction (or increase). All of these questions must be addressed in the protocol so that the treating clinicians know what to do in various circumstances.

Other information that may be relevant includes instructions on what to do if the patient is noncompliant in taking medication or if there are delays in treatment for reasons other than side effects. The latter could happen, for example, if the patient stopped taking study medication because of a short vacation. Would a gap of four days in the treatment plan be sufficient to make this case unevaluable? Questions like these will arise constantly throughout the trial, and since it is difficult to predict all possible situations that might arise, it is also important to include the name and telephone number of an individual to call should unexpected circumstances occur.

Schema

There is often a schematic diagram in the protocol summarizing the treatment plan in a pictorial format. If there is a schema, it should be completely consistent with the section of the protocol that describes the overall treatment plan. However, it should be emphasized that a schema is not a complete substitute for the treatment plan in the protocol and that treatment decisions should not be made based solely on the information in the schema. The schema is intended only to provide a pictorial overview. Figure 2.2 shows an example of a protocol schema.

Study Parameters

Most clinical trials require tests to be done at specific time points in the trial. The tests and schedule must be clearly defined in the protocol. This section usually includes information about tests that need to be done prior to entering a patient on study, both to ensure eligibility and to document baseline values/characteristics for the patient. Throughout the course of the trial, tests will be required to monitor patient safety and to assess response to the protocol treatment. It is important to be realistic in the requirements for the study, and to be sure that the tests required are clinically relevant and do not lead to excessive cost or inconvenience to the patient. Figure 2.3 gives an example of a study parameters section. The table also has notes that give information about relevant time frames for doing tests.

Data Collection Requirements

There should be a section of the protocol that details the schedule and type of data to be collected on the trial. Normally this will list the case report forms to be completed along with the relevant time schedule for their completion. If forms are being submitted to the Coordinating Center, the complete mailing address should be included. Figure 2.4 shows a sample "Records to be Kept?" section for an oncology protocol.

Statistical Considerations

As discussed earlier, the statistician for the trial should prepare a statistical design for the trial. This section should be detailed, contain information about the hypotheses that is being tested, the total accrual required for the trial, an estimate of the amount of time it will take to accrue those patients, and details of how the results will be analyzed and interpreted. There may be other information included depending on the type of study. For example, Phase II trials often have a sequential design where a number of patients are entered and assessed for response to therapy. If there have been a certain number of responses, additional patients are entered. If not, then the study is terminated. Phase III trials usually have stopping rules built into the design specifying when formal analyses will be done and under what circumstances the study should be terminated early. Details of these plans should be included in the statistical section. The section can also contain documentation on the statistical methodology that will be used in the analysis. If a high drop out rate or

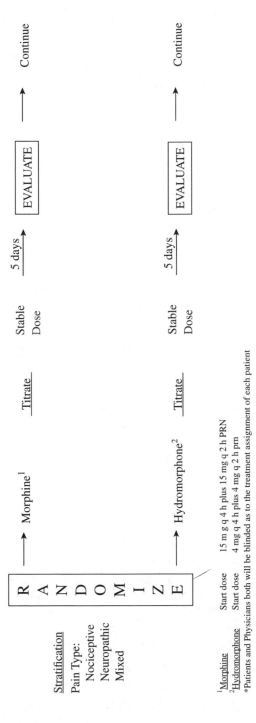

Figure 2.2. Sample Treatment Schema.

Stratification
Pain Type:
Nociceptive
Neuropathic
Mixed

[1]Morphine   Start dose   15 m g q 4 h plus 15 mg q 2 h PRN
[2]Hydromorphone   Start dose   4 mg q 4 h plus 4 mg q 2 h prn
*Patients and Physicians both will be blinded as to the treatment assignment of each patient

Morphine[1] Titrate Stable Dose 5 days EVALUATE Continue

Hydromorphone[2] Titrate Stable Dose 5 days EVALUATE Continue

STUDY PARAMETERS

1. All pre-study scans and X rays should be done ≤6 weeks before registration.

2. Pre-study CBC (with differential and platelet count) should be done ≤ 2 weeks before registration. If WBC or Plt counts are abnormal (see Section 3.0), they must be repeated <48 hours prior to registration.

3. All required pre-study chemistries, as outlined in Section 3.0, should be done ≤2 weeks before registration. If any required values are abnormal, they must be repeated <48 hours prior to registration.

When recording pre-study results on the case report forms, please make sure that ALL relevant dates are clearly given. Do NOT put all the results under the date for Day 1 of protocol treatment unless they were actually done that day. Record the actual dates.

	Pre-treatment	Day 1 Each Cycle (Every 4 Weeks)	Day 1 Each Cycle (Every 8 Weeks)	Off-Treatment Follow-up[4]
Physical examination, weight	X	X		X
Performance status	X	X		X
CBC with differential, platelets	X	X[1]		
Creatinine, AST, bilirubin	X	X		
Chest X ray	X	X[2]		X[2]
CT scan of chest and abdomen	X		X[2]	
Tumor measurement	X	X[3]	X[3]	X

[1]Perform weekly during the first two treatment cycles, then on Day 1 of each subsequent cycle. Counts may need to be checked weekly if Day 1 counts are low; see Section 5.312.

[2]If used as measurable disease parameter

[3]To be performed monthly if disease measurements based on physical examination or chest X ray; if measurements require other testing such as CT scans of chest and/or abdomen, perform tumor measurements every other cycle.

[4]Off-Treatment/Follow-up schedule:
- Every 3 months if patient is <2 years from study entry
- Every 6 months if patient is 2 to 5 years from study entry
- Every 12 months if patient is >5 years from study entry

Figure 2.3. Sample Study Parameter Section.

RECORDS TO BE KEPT

The following forms must be submitted to the Coordinating Center:

Form	To Be Submitted
On-Study Form	**Within 1 week of registration**
Flow Sheet	Baseline: within one week of registration
Measurement Form	On-Treatment: every month
	Off-Treatment: submitted according to the follow-up form off-treatment schedule below until relapse or disease progression
Follow-up Form Parts A, B, C, D, E	Every month while on treatment and at the completion of treatment
Parts A and B	Off-Treatment: • every 3 months if patient is <2 years from study entry • every 6 months if patient is 2 to 5 years from study entry • every 12 months if patient is >5 years from study entry
Toxicity and Complications Form	Every month while on treatment and at the completion of treatment
Adverse Event Form	Within 24 hours of an Adverse Event

Figure 2.4. Sample Records to Be Kept Section.

unevaluability rate is anticipated, allowances for that need to be built in to the sample size.

Measurement of Effect

To assess whether or not a treatment is effective, there must be criteria to measure the outcomes of the treatment. Since it is essential that each patient be assessed using the same criteria, these criteria should be specified in the protocol. In cancer trials, the end points being measured are usually survival, response to treatment (regression of tumor), and severity of side effects. To be able to accurately assess these outcomes, the protocol should include criteria for grading side effects and criteria to assess response. In a cardiac trial where the end point is length of time without symptoms, there should be a detailed description of the symptoms that would indicate failure; on an AIDS trial that uses a biologic marker to indicate the patient's disease status, there should be information about the values of the marker that indicate response and failure. Documenting the criteria for assessing outcome leads to consistent assessment for each patient entered on the clinical trial.

Definitions

It is useful to have a section in the protocol that provides clear definitions of terms that are critical to the protocol. For example, if a protocol requires a specific surgical procedure, details of that procedure should be included, or if a term is used that can be interpreted in different ways by different people, the term should be defined. A very simple example of the kind of term that could be interpreted differently is "elderly." If a protocol requires extra supportive care for "elderly" patients, then the relevant age group should be defined to ensure that there is consistent treatment for the elderly (and nonelderly) groups of patients entered.

Regulatory Requirements

Clinical trials should be conducted according to the laws and regulations of the country in which the trials are being done. Laws vary from one country to another, and it is important to be sure of the relevant regulations before activating a trial. The regulatory code generally referred to in the conduct of clinical trials is called the Code of Good Clinical Practice (or GCP), and this code defines procedures that must be followed to ensure data integrity and that the interests of the patient are being met. This critical aspect of the conduct of clinical trials is discussed in more detail in Chapter 9. If regulatory documents need to be collected during the course of the protocol, or special procedures need to be followed, this should be documented in the protocol. For example, if the protocol is being sponsored by a pharmaceutical company, and representatives of the company will be visiting each participating site to audit the accuracy of the data on the case report forms, this information can be included in the protocol.

Submission of Other Materials

Trials may require the submission of materials other than the case report forms that are required by protocol. For example, if there are scans or X rays that are to be read centrally to ensure consistency and objectivity in interpretation, there should be instructions in the protocol document on the procedures to follow in submitting these scans/X rays. Other trials may require the submission of pathology slides or blood/tissue samples. The instructions should include exact details of the materials to be shipped, when they are to be collected (e.g., pre-study, after one cycle of treatment, at time of response to treatment or at end of treatment), the method of shipment (e.g., whether special packaging is required, whether they should be sent regular or express mail), and the address where the materials have to be sent. Any special instructions on how the costs of shipping are to be covered should be included as well, as should any limits on the days of the week when materials can be *received* at the central repository and any delays in delivery that would affect the viability of the materials. For example, if blood/tissue is being sent and needs to be received and frozen within 24 hours of shipment from the participating site, but the laboratory receiving the samples cannot accept them on weekends or holidays, then the protocol must state that samples should not be shipped on Fridays or the day before a holiday.

Drug Ordering

If the protocol treatment involves drugs, there should be a section in the protocol that tells the participant how to get the drug. It may be commercially available from any pharmacy or it may have to be ordered specially from a drug repository for this trial. If relevant, a drug order form should be included as part of the forms set for the study, along with instructions on how and when to order. If the drug requires special handling or mixing procedures (e.g., mixing in a saline solution), this information should also be included. Figure 2.5 gives an example of a Drug Order Form.

Sample Consent Form

As part of the Code of Good Clinical Practice, patients must be fully informed about the trial and must consent to being entered on the study. There are usually national or local requirements that dictate the content and format of the information to be provided to the patient, and usually a consent form has to be signed by the patient and/or a witness prior to entry. In some countries the patient signs a detailed consent form that contains information on the patient's rights, the possible side effects of treatment, and information about whom to contact with questions. In other countries this detailed information is contained in a trial brochure to be given to the patient, but the actual consent form is fairly short. There are also countries where consent is given ver-

<table>
<tr><td colspan="3" align="center">Trial 0101
DRUG ORDER FORM</td></tr>
</table>

Requested By: _____   **Signature** _____

Patient ID _____   **Shipping Address** _____

_____

Note: No PO BOX Numbers

Patient ID	Initial Order (check if yes)	Reorder (check if yes)

FAX ORDER TO COORDINATING CENTER

Drug will be shipped (refrigerated) on Monday to Thursday for next day delivery. Drug can be shipped on Friday for Saturday delivery if site can guarantee receipt of Saturday delivery.

Shipment will go out on the day that the order is received.

Figure 2.5. Sample Drug Order Form.

bally. To ensure that necessary regulations are met, instructions for getting patient consent and a sample consent form can be included in the protocol. Chapter 9 discusses the consent process in more detail.

It is recommended that all drafts of the protocol be reviewed by the key personnel involved in the trial and by one or more of the investigators and Clinical Research Associates at the participating sites. Each will review the protocol from their own perspective and will ensure that the document is clear and consistent. Once the protocol activates, if problems are found with its contents, a formal amendment should be made to the document and circulated to all relevant parties, including the Sponsor. It is important that the protocol reflect the actual way that protocol patients are being treated, and except in cases of clinical necessity, exceptions should not be made to the protocol as written.

DATA COLLECTION SYSTEM

In parallel with the writing of the protocol document, the logistics of the data management system must be developed. Because the two are so closely inter-related, it is important that all key members of the trial team be involved in both and that there be constant iteration as both evolve. While this is a criti-cal aspect of managing large, complex multicenter trials, concurrent devel-opment of logistics will improve any trial. It means that the entire project must be considered in full detail and all aspects of trial design and data man-agement must be reviewed from different perspectives. The following is a summary of areas that require attention during the development and planning stages. Subsequent chapters will explain these fully.

Data Items to Be Collected

For the trial to be successful, it is essential to collect the data that will provide answers to the question(s) being asked. Keeping the volume of data within realistic limits is a difficult task, since there is always the tendency to collect more data rather than less, "just in case" these may be useful information. However, if participants are asked to do too much, the quality and complete-ness of the data can be compromised. It is better to focus on the key data needed to answer the objectives of the trial than to try to cover all possibili-ties. One approach to defining the required data is for the Statistician and Study Chair to develop a preliminary analysis plan for the trial, outlining the information that will be analyzed and reported. While it may seem strange to start at the end and work backward, it does force the team to think through the data requirements in a meaningful way.

Design of Case Report Forms

To ensure that the trial data are recorded in a consistent way for all cases entered, case report forms (CRFs) need to be designed and tested. The basic rules for forms design are the same whether paper forms or electronic screens are being used for data capture. Thought must be given to "when" and "where" data will be collected, and forms should be designed with this in mind. For example, if some baseline data are going to be collected by the radi-ology department and some by the cardiology unit, it would be better to have a separate form for each block of data rather than use one form that has parts to be completed by the separate departments. There are different designs for recording answers on forms, including the use of predefined code tables, mul-tiple choice formats, or free text with translation of responses into codes at the

Coordinating Center. Thought needs to be given to both ease of completion at the participating sites and ease of data entry from the forms. Once case report forms have been designed, it is recommended that they be piloted by one or more of the sites that will participate in the trial: This can provide valuable feedback on potential problems *prior* to the activation of the trial. Chapter 3 covers forms design in more detail.

DATABASE DESIGN

In parallel with the definition of data items to be collected and the development of the data collection forms, the trial database needs to be designed. These three activities are very closely interrelated, and all should be considered concurrently. The forms need to be designed with the database structure in mind. The database design should allow for easy retrieval of data both for the statistical analysis and for administrative purposes such as tabulating accrual by participating site. Depending on the complexity of the database structure, there may be multiple database records for each patient entered on the trial. If this is the case, then there needs to be a way to link all records for one patient, usually by including a unique patient identifier in each record. Decisions need to be made about whether all data should be computerized or whether some supporting data could be collected but not entered into the computer. For example, if a laboratory test is used to assess responses to the protocol treatment, all occurrences of this value need to be collected to be able to determine when a response first was observed, but it may not be necessary to enter all the test dates and results into the computer, only the date of the response. Database issues are discussed in more detail in Chapters 3 and 4.

PATIENT REGISTRATION

Before activating a trial, there needs to be a system in place for registering/randomizing the patients. It is strongly recommended that all cases be entered on the trial prior to starting protocol treatment. While this is essential for trials that involve randomization, it is also important in nonrandomized trials. There are two main reasons for this. First, it means that an eligibility check can be done at the time of registration to make sure that the patient meets all the entry requirements for the trial and that patient resources are not being wasted by entry of ineligible cases. A high number of ineligible cases on a trial will compromise the results and may require an increase in the sample size for the trial.

The second reason for prospective registration is to minimize bias in patient selection. For example, consider the situation where a clinician has two patients who are found to be eligible for a clinical trial. The clinician determines their eligibility on a Friday afternoon after the registration office has closed for the weekend. Although there is no urgency to begin treatment on this trial immediately, the physician decides to start the treatment, and plans to call and register the patients on Monday when the registration office is open. On Monday when the patients are seen again, one patient has tolerated the treatment well and is continuing to take the medication, but the second patient suffered severe side effects on Saturday after taking the medication and decided not to take anymore. Because the second patient is no longer taking the medication, the physician decides that it is not worthwhile entering this patient on the trial and only enters the first patient.

Scenarios like this introduce a bias in patient selection, because the side effects experienced by the second patient are critical data items to collect in assessing the overall benefits and drawbacks of the treatment. If this kind of selection process were followed by all physicians entering patients, the trial would include only patients who did not experience side effects. This would mean that there was no systematic way to evaluate the severity of side effects in the population under study, and if the medication was later made available to the general public, the incidence of side effects would be very much higher than that predicted by the results of the clinical trial. It is therefore recommended that all patients be formally registered before they start on treatment.

There are several ways to implement prospective registration, depending on the type of trial, the number and location of participants, and the resources available. Telephones, facsimiles, computers, and paper-based systems can all be used. If randomization is involved, the Statistician needs to define the algorithm to be used to randomly assign treatments. More details of registration/randomization systems are found in Chapter 6.

DATA COLLECTION MECHANISM

Decisions also need to be made on the mechanisms for ensuring that complete and consistent data are collected in a timely way. The decision about whether data will be collected on paper forms, or whether there will be computerized data collection at participating sites, is often based on the type of study, the number of sites participating in the trial, and the available resources. No matter how the data are collected, there needs to be a mechanism for transferring the data to the Coordinating Center for analysis. This can mean mailing paper forms, faxing forms, or transmitting data files electronically.

If paper case report forms are used, the forms must be designed and tested prior to the activation of the trial. A trial should never be activated without case report forms no matter how much pressure there is to begin patient accrual. If case report forms are not available, the investigators do not know what data to collect, and it is unlikely that all trial data would be available retrospectively for the patients entered on the trial before the forms were ready. If these patients were unevaluable for the study because key data items were missing, this could require an increase in the accrual goal for the trial to compensate for the unevaluable cases.

Participants need to have blank forms available when they enter a patient on the trial, so a mechanism must be developed for distributing copies of the forms. If there are only a few forms involved, they can be appended to the protocol, and participants can be required to make copies from this master set. If there are a lot of forms, or if the forms are being printed on bound multi-copy (no carbon required, or NCR) paper, then forms sets will need to be circulated separately and provisions made for renewing the stock at a site when existing ones are used.

There also should be a system developed to ensure that data are submitted in a timely way. While this kind of system is essential in multicenter trials, it is also important in a single institution study. If several patients are entered on the trial at the institution but forms are not completed as events happen, it is feasible that when the Clinical Research Associate eventually tries to complete the forms, there will be missing data for patients because no one reminded the physician that a certain key piece of information had to be collected. If the forms are filled out regularly, this is less likely to happen, since the missing data will be noticed earlier, and the Clinical Research Associate can then remind the clinician to collect it for future patients. For multicenter trials some kind of reminder system and method of monitoring data submission need to be in place. Finally there must be procedure manuals and documentation available for the participating sites to provide guidelines for completing the forms and meeting the data submission requirements.

If data collection is done electronically, with Direct Data Entry (DDE) at the participating sites, then data entry screens take the place of paper forms, and they need to be developed and tested prior to activation. Software is needed for doing consistency checks, range checks, and logical checks on data as they are entered so that possible errors can be corrected in real time. For example, for a trial requiring prospective registration, if the data entry operator enters a "Date Treatment Started" which is earlier than the "Date of Registration," there should be an error message and the user given the opportunity to correct the value at the time of entry. With electronic data entry, a system also needs to be developed for transferring the data to the Coordinating Center. Documentation for using the software and for entering the data must be written, and training must be available for new users. Systems also need to be put in place for ongoing user support after activation of the trial. More details about data entry systems can be found in Chapter 5.

REFERENCE CENTERS

If Reference Centers are being set up to do specialized reviews for the trial, the details need to be worked out during the development phase of the protocol. Reference Centers are usually used when certain materials are going to be reviewed centrally for a trial rather than at individual institutions. Examples of materials that might require central review are pathology slides, X rays, photographs, or blood specimens. A Reference Center provides specialized and consistent review procedures for the materials. Before the trial begins, the following need to be worked out:

- Are materials going to be shipped from the institution to the Reference Center, or are they going to pass through the Coordinating Center and be logged in?
- If they are going to the Reference Center directly, how will the Coordinating Center know what materials have been received?
- Who will be responsible for requesting overdue materials—the Coordinating Center or the Reference Center? If the Reference Center, a mechanism must be in place to notify the Reference Center when a patient is registered.
- Are there special shipping/packaging requirements for the materials?
- How are shipping costs to be paid?
- Do results of reviews have to be quickly relayed to the institution so that a treatment/clinical decision can be made? If so, how will the results be conveyed?
- How and when will the results be conveyed to the Coordinating Center?

Depending on the trial and the materials being reviewed, there may be other logistical details to work out. Communication between the Reference Centers and the Coordinating Center is very important, and as many details as can be predicted should be worked out ahead of time.

QUALITY CONTROL SYSTEMS

Participating Sites

The goal of a data collection system is to ensure timely collection of complete and consistent data and mechanisms for this need to be in place at the participating sites. For data to be complete, there needs to be a system in place for ensuring that study parameters are followed and that the tests required by the protocol are done according to the defined schedule. These systems are developed by the Clinical Research Associate at each site and can vary in approach and complexity. A copy of the parameters section from the protocol can be kept in the patient's medical chart as a way of reminding the clinician what tests need to be ordered. Alternatively, the Clinical Research Associate can check the protocol prior to the patient's visit, schedule any tests that need to be done, and provide a reminder note to the treating physician about any special questions that should be put to the patient. The Clinical Research Associate can also help with scheduling future clinic visits for the patient. It is often worthwhile to discuss the protocol requirements with the patient so that they are aware of the schedules and can help in ensuring that they are met.

To ensure consistent data over time, the Clinical Research Associate needs to have a mechanism for checking newly completed forms against forms previously submitted to make sure that data is consistent. For example, in a dermatology trial a patient may have several skin lesions, but only one is being followed to assess response to a new lotion that is being tested. It is important therefore that the data reported on response to the use of the lotion always be based on assessment of this indicator lesion.

To ensure the timeliness of data submission, the Clinical Research Associate at the participating site needs a mechanism for keeping track of when forms are due for a patient. Again, there are different ways of setting up such a system. Computers can be used to assist with tracking data schedules, but there are other ways that are equally effective, especially with small numbers of patients. For example, a wall/desk calendar can be used to track when data forms should be submitted. At the time one form is submitted for a patient, the Clinical Research Associate makes a note on the calendar on the date that the next form should be due. This note documents the form to be completed on that date, and by checking the calendar daily, the Clinical Research Associate can identify the forms to be done on that day. Similarly a system can be designed using postcards or 3×5 cards and a file box that has a section for each month of the year. When a patient goes on to the study, a card is filled out listing all the forms due for that protocol. A date is written against the first form due, and the card is then filed in the section for the month that corresponds with that date. Each month the Clinical Research Associate pulls the cards for the month and completes the forms that are marked. Each card is then updated with the date of the next form due and refiled in the appropriate section. These are simple but effective systems for helping to keep track of data due.

It is essential to ensure that correct patient identifiers are clearly written on all forms that are submitted for a patient on a trial, and the Clinical Research Associate should maintain a cross-referenced file of patient names and protocol identifiers. More details on local data management systems can be found in Chapter 7.

Coordinating Center

Quality control systems also need to be developed in the Coordinating Center. Before a trial is opened to patient entry, there need to be mechanisms for review and assessment of data which will be submitted. Documentation needs to be written for the procedures to be followed in evaluating the data and the steps to be taken if problems are found. For example, submitted data needs to be checked to ensure that the correct forms have been used, that patient identifiers are on each form, that data are consistent over time, and that the forms are complete. These checks can be done manually or by computer, and there needs to be a mechanism for sending queries back to the institutions when discrepancies are found. There also must be defined procedures for evaluating the study's end points for each patient, using the objective criteria in the protocol. If all data need to be reviewed by a clinician or another specialist as well as by the Data Coordinator in the Coordinating Center, then a review system has to be developed. This could involve copying all data as they are received and sending copies to the reviewer, or having the reviewer visit the Coordinating Center on a regular basis to review the data on site.

Conventions need to be developed and documented for flagging such things as missing data values, tests not done whose results are therefore not available, and cases with unresolved questions. It may be necessary to collect documentation that shows which sites are in compliance with the regulatory requirements for the trial. All these mechanisms must be in place before the study is activated.

The Coordinating Center can further develop tools to assist sites in complying with the protocol requirements. These could include generation of patient-specific calendars. When a patient is entered on the trial, a calendar can be generated to document the dates on which tests are to be done or data to be submitted. Calendars are most effective in studies where there are unlikely to be unpredictable delays in the administration of the protocol treatment. In studies where delays or changes are foreseen, such as when a patient suffers severe side effects, these calendars are less useful, since they can quickly become out of date. There can also be programs that use the database to generate listings of overdue data and queries about inconsistent data.

While all of these systems can be refined as the study progresses, it is important that they be designed and available prior to activation of the study so that equivalent quality control standards are applied to all data collected. More details on this can be found in Chapter 8.

It is important to ensure that data are collected and submitted in a timely way. If there is no mechanism to do this, there can be problems in interpreting the data that have been submitted. Often bad news comes in first, and the forms that are received at the Coordinating Center will document study failures. Unless there is a balance to ensure that data are received on all cases according to the same schedule, there could be a risk of over-reacting to the bad news and drawing erroneous conclusions about the efficacy of the treatments under study.

Reference Centers

It is important to know that the Reference Centers in a trial are producing consistent reviews of the trial materials and that they are maintaining adequate records for the trial. Quality control of the Reference Centers should therefore be built into the overall plan. This could involve site visits to the Reference Centers to review records, re-review of a sample of cases by other reviewers, or blinded re-review of some of the same cases by the reviewers in the Reference Center. The Statistician for the trial should be involved in developing this quality control process and in ensuring that the sample size of cases being re-reviewed is adequate to ascertain consistency and objectivity.

USE OF COMPUTERS

There are many ways in which computers can be used to assist with the conduct of a clinical trial, and several have been mentioned in the previous sections of this chapter. Computers can be used for the trial database, data collection/data entry, randomization/registration, and study analysis and in creating study management tools. In making the decision about the extent to which computers are used, functionality and cost should be assessed. There are several choices that have to be made about both hardware (the computers and peripheral devices) and software (the programs that run on the computers). Powerful personal computers (PCs) are now readily available at a reasonable cost and these can be used (either alone or networked) to manage many aspects of a clinical trial. If a more powerful system is needed, workstations or mainframe computers should be considered. When making a decision about the use of computers, consider the costs, functionality, and availability. If a system is already available for use, it may be necessary to use that system rather than purchase a new one. In this case the hardware choice is already made.

The same criteria apply to selection of software. It is essential to evaluate software packages to ensure that they will provide the functionality required at an acceptable price. It is recommended that wherever possible, commercial software be used rather than developing systems in-house. However, if commercial software provides only some of the necessary functions, it may be necessary to develop supplemental software. It is important that the database storage and retrieval system interface with the statistical software to be used for analysis.

If computers are used for the management of the trial—and for all but small Phase I and II trials, it is inconceivable nowadays not to use one—decisions have also to be made about whether all computing will be done in the Coordinating Center or whether parts of the system will be distributed to the participating sites. There are many factors that influence this decision, including the budget for the trial, the number of sites participating, the complexity of the data collection requirements, and the availability of qualified personnel. The decision to distribute some of the trial management to local computers should only be made after detailed analysis of resources and requirements. More information about the use of computers in clinical trials can be found in Chapters 4 and 5.

DATA-MONITORING COMMITTEE

Most Phase III trials are monitored by a special committee that is set up before the trial starts. The committee could be completely independent of the trial participants, or it could be a mix of participants and outside members. The Data-Monitoring Committee will be responsible for monitoring the progress of the trial until the data are unblinded. They will monitor the safety data from the trial and should have the authority to recommend suspension or termination if the side-effect rate is excessive either in number of events or in severity of events. The Data-Monitoring Committee will also be responsible for reviewing the trial when it reaches any of the interim analysis points stated in the statistical design section of the protocol. This committee is an important one and has a lot of responsibility for decisions about the trial. The membership of the committee may be decided by the sponsor or by the participants, but the committee should be set up prior to activation of the trial.

SUMMARY

This chapter has outlined many issues to be addressed prior to activating a clinical trial. For any trial, a requirements analysis should be done involving all members of the trial team. Once the requirements have been defined, the necessary systems need to be developed and be in place before patient entry begins. Obviously the needs are more complex for a large multicenter study than they are for a small study being done in one location; nevertheless, to ensure the integrity of the trial results and the best use of available resources, it is important that these requirements be considered even for small trials. A complete and well-written protocol document will ensure consistency over time for all patients entered, and adequate data management systems will allow the collection of complete and consistent data in a timely way. Computers can assist in many aspects of trial management. Errors or omissions at this stage of the trial can be very costly and can lead to a worthless trial. After a careful review of the study requirements, appropriate decisions should be made based on available resources.

CHAPTER 3

Data Definition, Forms, and Database Design

The identification of the data items to be collected for the trial, the subsequent design of the data collection instruments, and the design of the computer database are critical to the success of a trial. These three activities are interrelated and are best done in the order listed. That is, first comes the identification of the data items to be collected for the trial, and then the design of the data collection instruments and the database. It is always useful to look at the data collected for other similar trials before designing new forms. If there are existing forms that can be used for the new trial, this eliminates a lot of extra work. For this purpose there is a book with more than 600 forms that have been used in clinical trials, and it provides plenty of examples that can be used for new trials.[1] If such existing forms are used, it is important that the forms collect only the necessary data for the new trial and not any superfluous data. The data collection instruments, whether on paper or electronic screen, should be available prior to entry of the first patient on the trial. A trial should not be activated without data collection instruments, regardless of how much pressure there is to start accrual.

DEFINING DATA ITEMS TO BE COLLECTED

There are different types of data that may need to be collected for a trial, and it is important during the planning phase of a study to think through all the requirements for the trial. For example, as well as collecting the research data, it may also be necessary to collect data to help with the administration of the trial and also data to document compliance with regulations and Good Clinical Practice (GCP).

Identification Data

When forms are submitted for a trial, it is essential that they be linked to the appropriate patient and also to the correct trial. Often one Coordinating Center is responsible for the collection of data on many clinical trials and receives completed case report forms for all of them. A form must therefore show a space for recording sufficient information for correct identification. These identification data items should always include a patient identifier and, unless it is already pre-printed on the forms, a trial identifier. It may further be useful to collect the name of the institution that submits the data in case there are errors in other identification data. Knowing the institution often allows errors to be detected.

Research Data

Research data provide the information that is analyzed to answer the questions being asked in the study objectives. The identification of these data items should be done during the protocol development phase and should involve all members of the trial team to ensure that they all have input.

It is always tempting to collect data items "just in case" they turn out to be interesting when the data are analyzed. However, collecting too much data can be detrimental because, as the volume of data required increases, the quality of the data recorded on case report forms decreases. It is therefore important to limit data collection to those items which are truly necessary to answer the trial objectives and manage the trial. In assessing the data requirements, the team should distinguish between data needed for the clinical care of the patient and the data needed to answer the research objectives. Data collection for the trial should be limited to the research data. For example, to ensure that the patient has appropriate clinical care, it may be important to know exactly when a medication was given, but this level of detail may not be important when analyzing the data for the trial. The information will be recorded in the patient's medical record, but there is no need to collect it on the data forms for the trial. In many trials only a small fraction of the clinically relevant information will end up as part of the trial database.

Omissions at this stage are almost impossible to overcome once the trial has begun to accrue patients, since it is hard to collect data retrospectively. If specific information was not collected and recorded in the patient's record, it is lost to the trial. The data will therefore be missing for every patient entered before the omission was detected and the data collection forms were changed to include it. It is even worse if the missing data are not discovered until the time of study analysis, for then the data will be missing for every patient entered.

To avoid this, during the development stages the Statistician and Study Chair should prepare a preliminary analysis plan outlining the information that will be included in the final report of the trial. The other members of the trial team can review this outline and provide input. Once this has been done, it will be easier to identify the data items that need to be available for eventual analysis. At minimum, this will usually include key dates of events such as date of entry onto the trial, information about patient safety while on the trial if side effects of treatment are expected, and information on the study's end points for each patient.

Administrative Data

In addition to the data required to answer the research questions, it is often necessary to collect administrative data to help with the management of the trial. The amount of administrative data depends on the size of the trial and complexity of the trial structure. In a small single institution trial, much less information is needed than in a large multicenter trial where data and materials are being shipped to various locations.

During the trial the Coordinating Center must be able to contact participating sites to request missing data as well as to ask for clarification on submitted data. This means that when a patient is entered on a trial the Coordinating Center must register the name of the participating site. It may also be necessary to collect information about the treating physician and the person who is responsible for data submission. This information facilitates communication between the two sites.

If the trial involves shipping of materials to Reference Centers, then a tracking system is needed. The information collected should indicate what materials were sent, when they were sent, and where they were sent. The receipt of the materials also is recorded. Other administrative data depend on the specifics of the trial.

Regulatory Data

Some trials may call for documentation that shows compliance with regulations. Such documentation could include protocol approval by an Ethics Committee/Institution Review Board prior to patient entry, consent of the patient prior to entry, and qualifications of the personnel at a participating site. In some instances, instead of copies of the documents, it may be acceptable to only record that the institution confirmed that the documents exist. The requirements should be defined prior to the start of the trial. Normally, if the trial includes investigational treatments, copies of the actual documents are collected, and during the trial it may be necessary to collect additional or updated regulatory data. More information about regulatory requirements can be found in Chapter 9.

Reference Center Data

If trial materials are being reviewed at a Reference Center, it is important to define the data that will be generated as a result of the review and to decide how the data will be communicated to the institutions and to the Coordinating Center. There may also be administrative or quality control data that need to be collected at the Reference Centers, such as inventories of materials received and dates of receipt, and results of routine checks on the accuracy of the techniques being used for review. The Coordinating Center should ensure that all these requirements are defined before the trial activates.

DESIGN OF CASE REPORT FORMS

Once the data items have been defined, case report forms (CRFs) are developed. A case report form is a printed or electronic document that is designed to collect the required research, administrative, and regulatory data for a clinical trial. The measurement and recording of the trial data are perhaps the most critical steps in the overall data management process, and it is therefore important that the CRFs be designed with clarity and ease of use in mind. The design of CRFs has a direct impact on the quality of the data collected for a trial, so it is worthwhile to take time over the design and development of the forms and to develop a layout that is user-friendly. Case report forms should always be available before a trial is activated. Activating a trial without the CRFs in place is likely to result in a trial with incomplete and inconsistent data. Therefore the urgency to activate a trial should always be balanced by the need to have these important tools in place. It is recommended that forms be piloted by some of the trial participants prior to activation of a trial. The piloting can be done by completing the proposed forms using historical data from appropriate medical records available at the sites. This process allows the eventual users of the forms to have meaningful input into the design and piloting the use of the forms can often identify problems that can be corrected prior to implementation of the forms in real time.

The data collection forms should be concise and collect only the necessary data. When designing forms for a trial, thought should be given to the following.

Content and Organization of Case Report Forms

The ultimate objective of the CRF is to collect data that answer the trial's objectives. Once the required data items have been identified, it is necessary to decide how to organize these data items on the forms. When designing forms, keep in mind that it is not always best to minimize the number of forms by trying to fit as much as possible onto one page. It may be better to have more forms, each with a small amount of data. When deciding on the forms that are needed for a trial, ask yourself:

1. Who will be completing the forms?
2. When will data be available?
3. Where will the data be collected?

As a first step, the **timing** of collection of the different items should be established. For example, identify all of the data items to be collected at the time that the patient is entered on the study. This normally includes data on the patient's past medical history data that confirms the patient's eligibility and results of baseline tests required by the protocol. Other relevant time points could be the data collected during all or part of the protocol treatment period, the data collected when a patient finishes treatment, and any data collected as follow-up after the treatment period. All of these are logical divisions and can help in deciding which data items belong on which forms.

Besides the timing of the data collection, it is also useful to identify **where** the data will be collected and **by whom.** These are also logical divisions that can help to decide which data should be collected on which form. For example, there may be baseline data gathered from the medical record by a Clinical Research Associate and other data completed by a medical specialist such as a surgeon. Even though all the data are collected at the same time point, it would be more efficient to have two different forms for recording the information—one for the Clinical Research Associate and one for the surgeon. This allows the two people to complete their parts of the data collection requirements in parallel, rather than one person having to wait for the other to complete their section of the form before passing it on. Likewise, if some of the data is available in the cardiology department and some in the physical therapy department, two separate forms may work best.

These steps will help to organize the data items into logical groups of items, and individual case report forms can be created using these groups.

Format of Questions and Coding Conventions

The purpose of case report forms is to collect complete and unambiguous data for the clinical trial and to ensure standardization and consistency of data across participating sites. Their format should be designed with three functions in mind:

- Recording data on the form
- Data entry into the computer
- Data retrieval for analysis

The person recording data on the form should be able to answer the questions and record the answers in an unambiguous and efficient way, minimizing any possibility of misinterpretation or transcription errors; the person who is entering the data into the computer should be able to transcribe values from the form to the keyboard with minimum effort in following the flow of data and entering the data values; the person who is analyzing the data must be able to interface the data and the statistical software with a minimum of data conversion. Even if the data are not computerized but tabulated manually, it is important to design the forms with analysis in mind.

There are several ways to phrase questions on case report forms, but there are conflicting ideas on which is most effective in collecting complete, accurate data. Since the goals of different trials vary and the environments in which data are collected can be very different, it is recommended that the designer review other forms that have been used in clinical trials and adapt a format that suits the research being done and the resources available for a particular trial. If several forms are developed for a trial, the most important criterion is to use a consistent design across forms so that the users can become familiar with the format used. Page layouts should be similar across forms, and the headers of the pages should be designed in the same way, collecting the same identification information. Coding conventions should also be consistent for all data items. For example, 1 = no, 2 = yes for all no–yes possible answers.

The questions asked on forms can call for different types of responses. Some may be text strings, such as the name of treating physician; others may be categorical values reflecting results of laboratory tests; still others may depend on patient characteristics and require the person completing the form to select the appropriate answer from a list of possible choices.

Lengthy text strings are usually restricted to information that will not be analyzed directly. For example, a form could require the completion of a coded variable indicating whether the patient was taking antibiotics and then, if in the affirmative, ask for the names of the antibiotics to be written on the form. In the analysis, the Statistician can easily use the coded variable indicating that the patient was taking antibiotics, but unless the text string with the antibiotic names was subsequently translated into codes, it would not be easy to do any analysis on the actual types of drugs being taken. If text strings are collected, the form should allow adequate space for handwriting the information in one or more lines on the form, thus allowing the user to enter free text. Asking the user to put each character in a separate box can lead to frustration and incomplete information being entered.

When collecting categorical values, the important considerations are to provide the correct number of boxes for the answer; to preprint any required decimal points, commas, or other punctuation; and, when relevant, to specify the units to be used in recording the data. For example, in collecting the patient's weight, the following formats would be possible:

1. |__|__|__| lbs
2. |__|__|__| lbs |__|__| oz
3. |__|__|__| kg
4. |__|__|__| . |__| kg

If format 1 or 3 is used, the instructions for the form should indicate when the user should round the value up or down to the nearest whole number.

A common type of question on a form is one where the user has to select the correct answer from a list of given values. The following are some possible formats for these types of questions:

1. *Multiple Choice Format.* For a specific question all possible answers are displayed on the form, and the user has to circle/check the correct answer. This format is usually used if optical scanning is going to be used to convert the answers into electronic format.

   EXAMPLE:

   What is the gender of the patient?   (Circle one)   Male   Female   Unknown

   If a multiple choice format is used, the instructions for answering should be consistent across all questions, and the user should be asked either to circle the correct answers or to check a box beside the correct answers. Mixing the two ways of indicating the answer should be avoided. Multiple choice questions can be convenient for the person completing the form, but unless the answers are being scanned electronically, they are less conducive to efficient data entry. Normally, as the answers are being scanned, they are automatically converted to codes, and these codes are stored in the database. If the data are being entered by a data entry operator, a Data Coordinator may have to translate the answers to codes prior to entry so that the operator can more easily enter the data. The multiple-choice format allows easy completion of the forms but may require programming or Data Coordinator effort to convert answers to an electronic format, and because answers are converted to codes, it has no negative effect on the ease of analysis.

2. *Self-Coding Forms—Numeric Codes.* All possible answers to a question are displayed on the form, and each answer is assigned a numeric code. There is a space or box(es) where the answer is to be entered, and the user is required to enter the code corresponding to the correct answer.

EXAMPLE:

_____ What is the gender of the patient?

    1. Male

    2. Female

    9. Unknown

The user enters "1," "2," or "9" in the space provided depending on the patient's gender. This format is less intuitive for the person completing the form. The codes themselves are meaningless in isolation and only relevant when linked with a specific question and answer. However, when using this format, data entry is simplified and data analysis can be done using the actual values recorded on the forms. Note that the code for unknown is given here as '9', and not '3'. It is good practice to assign the same value for 'unknown' across all variables. As this is a single digit answer and other questions may require more than 3 possible answers, it is best to assign '9' as the consistent code for 'unknown' for all single digit fields. In this way, 'unknown' will always be the last possible answer after all other possibilities have been reviewed and considered.

3. *Self-Coding Forms—Nonnumeric Codes.* All possible answers to a question are displayed on the form, and each answer is assigned a code that is more meaningful than a randomly assigned numeric code. There

is a space or box(es) where the answer is to be entered, and the user is required to enter the code corresponding to the correct answer.

EXAMPLE:

_____ What is the gender of the patient?

    m—Male

    f—Female

    u—Unknown

The user enters "m," "f," or "u" in the space provided. This format is similar to the one using numeric codes, but here the possible answers are more intuitive to the person completing the form because the code being used is meaningful. Data entry can be slower using these codes, since the operator has to use the full keyboard for entry rather than just the numeric pad. With most modern statistical packages, analysis can be done using the actual codes recorded, but some packages may require conversion to numeric codes.

Whatever format is chosen for this type of question and answer, it is recommended that it be consistently used on all forms for a study and that, overall, only a small number of item formats be used.

Format of Questions

When formulating the questions which will go on the forms, these guidelines will help in achieving clarity:

1. Keep the text of the question as short as possible. The question does not have to be posed as a complete sentence if a short phrase is sufficiently clear.

   EXAMPLE:

   __ • __°F—Temperature on cycle 1, day 5

   This is more concise and just as clear as

   __ • __ °F—What is the patient's temperature on the fifth day of the first cycle?

2. Use terminology that will be familiar to the person completing the form. While words and phrases can be very familiar to members of the trial team designing the form, they may not be as clear to the Clinical Research Associates who have to fill out the forms in the different insti-

tutions. Piloting draft forms can ensure that the meaning of the questions is clear and that the required data will be collected.

3. Ask only one question at a time and do not introduce compound questions that can be confusing.

EXAMPLE:

_____ Is the patient fully ambulatory and on a regular diet?

    y—Yes
    n—No

This is really two questions, and only if the answer is "y" can it be fully interpreted. If the answer is "n," the patient might:

1. be fully ambulatory but not on a regular diet
2. be on a regular diet but not fully ambulatory
3. be neither ambulatory nor on a regular diet

If it is important to know the exact answers for both, then two separate questions should be asked:

_____ Is the patient fully ambulatory?

    y—Yes
    n—No

_____ Is the patient on a regular diet?

    y—Yes
    n—No

4. Instructions should be positive rather than negative. That is, tell the person completing the form what to do rather than what not to do.

EXAMPLE:

Answer all the following questions

Rather than

Do not skip any of the following questions

Consistent coding conventions are also advisable when designing case report forms. The user becomes familiar with one way of completing a certain type of field and is more likely to provide the correct answer if that type of data is always collected in the same way.

Dates. When dates are recorded, it is important to clearly identify the format to be used. In the United States, dates are usually recorded as month, day, year, but in many other countries, dates are recorded as day, month, year. Therefore, to reduce ambiguity and ensure accurate data, the forms should clearly state which format is expected:

EXAMPLE:

_/ _/ __ Date of birth

mm dd yy

In trials with international participation, it may be less confusing to abbreviate the name of the month and use that instead of the numeric representation. For example, 10–dec–95 rather than 10–12–95 (day, month, year) or 12–10–95 (month, day, year). Also, to deal with the turn of the century, it is wise to include century in all date formats.

Decimal Points. For answers that require the entry of a decimal point, the placement of the decimal point should be preprinted on the form so that the interpretation of the value entered is unambiguous.

Units of Measurement. When applicable, the appropriate units of measurement should be preprinted on a case report form. If there is a possibility that between participating sites, results of a test may routinely be reported in different units, then either the forms should allow for reporting in varying units or the documentation provided to the site should include a conversion algorithm. In general, it is advisable to collect the actual value and units from the site and do all conversions centrally to ensure accuracy. It may also be necessary to collect normal ranges of laboratory test results, since these can also differ across laboratories and, without the ranges, it is impossible to know whether or not a result is normal.

Unknown/Not Applicable/Not Available/Not Done. There should be a consistent convention for recording these categories. It is sometimes important to be able to distinguish between a test that was not done and one where the results are not available at the time the form is completed. If a test was not done, the Coordinating Center will know that it is futile to request that the missing data be provided. However, if the test was done but the result is not yet available, then a query at a later date will usually recover the missing information. Whenever a question could have one or more of these categories as a feasible response, a code should be provided for each relevant category.

Other. It is often difficult to predict all possible responses to a particular question, and in these circumstances it is advisable to allow for "Other" as a possible response. If used, there should also be provision for collecting the "Other" answer.

EXAMPLE:

_____ Give reason for stopping treatment

   1. Completed per protocol

   2. Patient refused to continue

   3. Serious side effects

   4. Progression of disease

   5. Other, specify _____

   9. Unknown

If the response to this question is that drug was unavailable at the time the patient was to be treated, then the person completing the form would fill in code 5 and in the space after "specify" would write that "drug was unavailable." In the Coordinating Center where data are reviewed, repeated use of "Other" should be monitored, and if a single reason keeps occurring, then either a code for that reason should be developed internally so that it can be recorded in the database, or consideration should be given to revising the form and adding that reason with a specific code. In this way the Statistician will be able to easily include this in the analysis.

LAYOUT OF CASE REPORT FORMS

As well as having a consistent format for questions on a form, there are several principles which should be considered in deciding how to organize the questions on a page or a screen. It is important that the form be legible in the conditions under which it will be completed. The print should be large enough to read easily (minimum of 8-point type for text), and if boxes are being used to collect coded answers, they must be large enough to write in. The layout of the questions on a form should be visually pleasing and should allow for ease of data entry.

Use of different fonts, bolding, italics, and underline can help to highlight areas of a form but only if used sparingly. If there are many different formats for text on the form, the users will only get confused and will not realize what is being brought to their attention and what is not.

To make sure that the response to a question is written in the correct place, the response field should be located close to the specific question. This means that the response field should be either at the beginning of the question or at the end. Consider the following examples:

EXAMPLE 1: Responses to the right of the question:

Patient's age at time of registration   ————

Gender (m = male, f = female)   ————

Race (1 = Caucasian, 2 = black, 3 = Hispanic, 4 = other,

    9 = unknown) ————

Date of diagnosis __/ __/ __

           mm dd yy

EXAMPLE 2: Responses immediately to the left of the question:

_____ Patient's age at time of registration

_____ Gender/(m = male, f = female)

_____ Race (1 = Caucasian, 2 = black, 3 = Hispanic, 4 = other,

    9 = unknown)

__/ __/ __ Date of Diagnosis

mm dd yy

EXAMPLE 3: Response aligned on the right:

Patient's age at time of registration .... _____

Gender (m = male, f = female) .... _____

Race (1 = Caucasian, 2 = black, 3 = Hispanic,

4 = other, 9 = unknown) .... _____

Date of diagnosis ....__/__/__

                                  mm dd yy

EXAMPLE 4: Questions right-justified followed by aligned answer fields:

           Patient's age at time of registration _____

                  Gender (m = male, f = female) _____

       Race (1 = Caucasian, 2 = black, 3 = Hispanic, _____

                    4 = other, 9 = unknown)

                 Date of diagnosis __/ __/ __/

                       mm dd yy

Example 1 has the questions aligned on the left-hand margin with the response field at the end. Persons completing the form read the question and then fill out the answer without having to shift the direction of their vision. However, the answers are buried in among the text and are not easy to follow when data entry is being done. The data entry operator will have to scan the question before finding the answer.

Example 2 has the responses aligned in the left-hand margin, and data entry can be done by reading down that margin without having to read the actual questions. This second version will probably take users slightly longer to complete because they have to read the question from left to right and then return their vision to the left margin to fill in the correct answer.

Example 3 has the questions aligned on the left margin, and dots used as fillers to align the response fields on the right. While this does make data entry easier than in Example 2, and does provide some help in linking the question to the correct response field, the fillers may have to be manually tracked across the page when questions are short.

Example 4 has the right margin of the question aligned, with the answer fields immediately to the right. This means that the user can read the question and fill in the answer fairly easily and answers are aligned for ease of data entry. This format is good where questions are short, there are minimal instructions, and the list of possible answers is short. For forms where there are long questions or a lot of instructions, it is less appropriate.

The format in Example 2 is recommended to improve accuracy both in recording the answer and in entering the responses into the computer. Format 4 is also appropriate when the questions and instructions are short.

LAYOUT OF PAGE

Usually case report forms can be designed in either single-column or double-column format, and there is no indication that one leads to better data than the other. Clearly, having two columns of data and less 'free' space on the page allows more data to be recorded on one sheet of paper. However, the resulting form may look cluttered and confusing. Cutting and pasting sample pages in both formats should give an indication of the best layout for a particular form.

In some trials, the two column format has been used with questions on the left-hand side of the page and arrows leading to boxes with explanatory text or conditional questions on the right-hand side.

Example:

Was treatment given according to protocol?

    1   Yes

    2   No

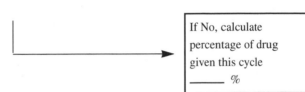

If No, calculate percentage of drug given this cycle _____ %

Clearly there are many different ways to lay out questions on a form, and the methods chosen will depend on the type and complexity of data being collected and on the background of the person who will complete the form. It is important that the form is easy to follow and that the instructions are clear. Keeping a reasonable amount of white space (space with nothing in it) on a form often makes it more pleasing to the eye.

HEADER

A standard header format should be developed for all the case report forms for a trial. Under the name of the form, the header should collect the data necessary to identify the patient and the trial. It is also useful to record the name of the institution that entered the patient. In any trial it is essential to be able to uniquely identify each patient entered and to ensure that all data for that patient can be linked during analysis. A unique patient identifier should therefore be assigned to each patient entered on the trial and this patient identifier should be recorded on all forms/materials for that patient. The identifier can be a unique number assigned to the patient at the time of entry, or it can be the patient's hospital ID number, the patient's initials, or a combination of these. It is essential that the identifier be unique within the trial and that no other patient entered on the trial can have the same identifier either by chance or design. This is one reason why patient names are not recommended as identifiers. Apart from the issues of patient confidentiality, it is certainly possible that two patients with the same name can be entered on the trial. Once the format of the identifier has been decided, it should be collected on every page of all case report forms for the trial, although the additional identification information only needs to be collected on the first page of a multiple-page form.

For some case report forms it may be useful to leave blank space on the form so that the person completing the form can write additional comments. Often such comments are useful to those who are reviewing the data. If comments or free text are collected, the Coordinating Center must review all that is written when they do quality control, since the information may need to be converted into computer codes and entered into the database. For example, a comment may provide information that allows more accurate grading of a side effect.

NUMBERING DATA ITEMS ON FORMS

Often each data item on a form is assigned a unique number for easy identification of that specific item. The numbering can be unique within the form or unique within all forms for the study. This is convenient for the Coordinating Center when requesting clarification for specific items. The item can be identified in a query letter or error report by its number. The participating site can then check the form to find the item in question.

ELECTRONIC SCREENS

While the guidelines given here have been primarily aimed at the design and development of paper case report forms, many of them also apply to electronic screens designed for direct data entry. Electronic screens have size limitations compared to paper forms. Most screens display 24 lines of text, and it is expensive to replace these with larger monitors that allow the display of a full page of data. However, when designing screens for data entry from case report forms, it is beneficial to have the screen format resemble the paper format as closely as possible. If data are entered directly from source without intermediate transcription to forms, then it is important for the screens to be as easy to use as paper forms, and many of the same design guidelines apply.

PRINTING AND DISTRIBUTION OF FORMS

Once the forms design is complete, decisions need to be made about how to print and distribute the blank forms to participants.

During the trial there will normally be a need for more than the original copy of the data to be maintained. For example, if data are being submitted to a Coordinating Center for review, the original of the form should be submitted to the Coordinating Center and a copy retained at the participating site. This gives the site a copy for reference if queries about the forms are sent by the Coordinating Center. There may also be a need to make a copy for the Study Chair, a copy of some forms for the reference labs, or a copy for on-site monitoring visits. The number of copies needed will vary from trial to trial and should be specified in the protocol or the Procedures Manual for the trial.

If only one or two copies are needed in addition to the original, it may be appropriate to print the blank case report forms on multi-part NCR (No Carbon Required) paper. This is specially treated paper that has ink capsules built into the surface of all copies except the top one. The NCR pages are glued together along one margin, and as the user writes on the top copy, the pressure causes the writing to go through to the other copies. Each copy can be a different color, and distribution of copies can be by color. For example, the white copy to the Coordinating Center, yellow copy to the study chair, and pink copy to be retained at the site. The advantage of NCR paper is the reduction in time and expense of making copies. However, if more than two copies are needed in addition to the original, heavy pressure needs to be applied when filling out the top form, and this can slow down the completion of the form. The quality of copies beyond two copies is still likely to be poor, and also the quality of the copies deteriorates over time. Other disadvantages are that the blank forms must be printed—they cannot be copied from a master form on a copy machine.

If the forms are designed with different parts of the forms printed in different colors of ink, then the forms must be printed or copied on a color photocopy machine. For the majority of clinical trials, this would be too expensive to be viable, and there are no clear data indicating that varying color on a form leads to better data. If color is used on a form and is an integral part of the instructions for completion of the form, (i.e., different instructions apply to sections printed in different colors of ink), then all forms must be printed/copied centrally and sent to the participating sites. It is unlikely that the sites will have facilities for making their own copies.

Forms can also be printed/copied on colored paper to clarify instructions for completion. For example, if the same forms are submitted twice during a study, the instructions can ask the participant to submit the blue forms for cycle 1 and the yellow forms for cycle 2. If the instructions depend on the correct colors being used, then again the forms should be printed centrally and distributed.

The majority of clinical trials use white paper with black print. The forms can then be printed/copied centrally, or one copy can be sent to each participating site and local copies made from that copy.

Once forms are printed, blank sets need to be distributed to the participating sites. There are several ways of doing this. If the number of pages is small, and white paper with black ink is used, the forms can be appended to the protocol document and distributed to the participants with their copies of the protocol. At the sites a master copy can be kept, and additional copies made as patients are registered to the trial.

If the forms are specialized (e.g., NCR forms, color print, color paper) or if the set of forms for one patient is large, the sets can be prepared centrally and mailed to the sites as needed. A decision will need to be made about whether one set is sent to the site and replaced when a patient is entered, or whether multiple sets will be provided at the start of the study, with the provision to reorder when additional sets are needed. Often the forms are made into booklets for each patient, in a loose leaf or bound format. With booklets, if the same forms are to be filled out at multiple times during the study, copies should be included for each relevant time point. This ensures that the participants follow the completion schedule but can lead to wasted paper, since not all forms will be completed for every patient. For example, if a patient goes off-study after the first dose of a study drug instead of remaining on treatment for 12 months, there are likely to be many pages in that form booklet that do not get used.

Decisions on printing and distribution should be made after reviewing the organization for the specific trial and the resources available. Forms should not be printed or distributed until the protocol document is finalized and all members of the trial team have approved the final forms.

MODIFICATIONS TO FORMS

Modifying case report forms after activation of a trial should be done only when absolutely essential. If it is necessary to make changes, the following should be considered:

1. If the change adds a question, consideration should be given to its location on the form. If the data entry screens and database records are exact images of the forms, adding the change at the end of the form will simplify the changes that will need to be made to them. However, separating the new data item from other related fields on the form may make it more difficult for the person completing the form. If the new question is added in the middle of the form, then care must be taken in restructuring the data entry screens and the database to accommodate the

change. If both old and new versions of the forms are still being submitted, allowance needs to be made for that. Other existing software may also need to be modified.

2. If the change adds another possible answer to the list of options for a particular question, and the answers are recorded as numeric codes, then the next unused code number in sequence should be assigned for the new option. It should not be inserted in the middle of existing options, and those options renumbered. For example, if the first version of the form used codes 1, 2, and 3 for options, the new option should be code 4, even if the option would be more logically be inserted after option 2. Using code 3 to mean one thing on the first version of the form and something else on the second version will cause errors and confusion in the analysis.

3. If questions are being removed from the form, extreme care must be taken to ensure that the subsequent electronic records are compatible with the ones created prior to the change. If a field is deleted on the form, but kept in the computer records which were generated from the original form, then allowance for the blank space must be made so that a location in the database does not have two meanings depending on the form used.

When forms are developed, it is advisable to have a form identifier and a version number on them. The date of creation is also useful. When modifications are made to the form, the version number and date should be updated. This can help to eliminate confusion and ensure that the correct version of a form is being used.

DATABASE DESIGN

The design of the database for the trial is closely related to the definition of data items to be collected and the design of the case report forms. The maximum amount of data that can be entered into the database is the complete set of data collected at the sites and submitted on the case report forms. However, for many trials there is no need to actually computerize every value. Sometimes a series of data items can be summarized into one overall value. For example, if the analysis will only use the worst grade of each reported side effect, then instead of computerizing all occurrences of that particular side effect, it may only be necessary to enter the worst grade into the computer. Similarly there may be data collected to verify a key variable. For example, there may be several chest X rays done to assess whether a cancer tumor has responded to chemotherapy. The measurements taken from these serial chest X rays are all reported, but in fact the only data that may need to be entered into the computer is the best response to treatment for that patient. The measurements ensure that the participating site is reporting the response to treatment correctly, but the actual values may not be important for analysis.

By scrutinizing the data being collected and assessing how it will be used, decisions can be made about the content of the trial database. Once this has been done, the structure needs to be designed. Chapter 4 discusses the kinds of software that can be used to maintain a trial database, and the software selected will have an impact on the structure of the database.

A commonly used database structure will have records that mirror the case report forms. For each type of form used in the trial, there will be a database record. With this structure it is essential that all records for one patient can be linked together. The unique patient identifier for the trial should be part of each record so that this linkage is possible. If the same form is submitted more than once for the same patient, then there also needs to be a way to distinguish between the different versions. Perhaps a date can be used to make the records unique, or a sequence number or visit number can be assigned to each. If there is a possibility that not all forms will be received for each patient, the statistical software needs to be able to deal with missing records. If the software is not able to handle missing records, then dummy records may need to be inserted to ensure complete records for all patients.

If multiple trials are being done that collect some data common across studies, the database design can allow storage of the common data in the same area, regardless of the data collection format. If the database is a large one, with many different sources of data and many different record types, then the database should be designed and set up by a qualified systems analyst/database administrator. Poorly structured databases can be very inefficient in terms of storage and retrieval, and errors can easily be introduced if the linkage between records is not adequately defined. However, for small studies done in single locations, the capabilities of microcomputer software packages are usually adequate for database storage and retrieval.

SUMMARY

Deciding what data to collect, designing and printing case report forms, and developing a computer database for a trial are critical for success of a trial. Adequate time should be allowed for these important steps, and the entire trial team should have input into all three. Errors during this stage of planning can be very costly for a variety of reasons. Failing to collect important data items, collecting too much data and compromising the quality of essential data items, designing forms that are unclear and difficult to use leading to errors in transcription and recording, or building a database that is inefficient are all examples of problems that can jeopardize a trial. Case report forms should be piloted prior to their introduction for a trial, and a trial should never be activated without data collection instruments in place.

REFERENCE

1. Spilker B, Schoenfelder J. *Data Collection Forms in Clinical Trials,* Raven Press, New York, 1991.

CHAPTER 4

Computers in Clinical Trials: Hardware, Operating Systems, and Database Management Systems

There are many ways in which computers can be used to facilitate the conduct of a clinical trial. They can be used for data entry, database storage, and management and for statistical analysis. They can be used at the coordinating center, the participating sites, and ancillary reference centers. The more locations that are involved, the more complex the system becomes. Software tools can be developed to assist with the conduct of a trial and to ensure that all trial requirements are being met. This chapter outlines some of the ways in which computers can be used and also suggests some issues that need to be considered before making decisions either about hardware or software. Chapters 2 and 3 also contain relevant information about the use of computers in clinical trials.

On computers, there are three main decisions to be made:

- Hardware
- Operating system
- Software

All three are interrelated, and a decision made about any one of the three will have implications for the other two. In some instances all three may have to be purchased, while in other instances there may already be existing resources that have to be utilized, thus narrowing any choices to be made. The possible options for all three should be researched together, and decisions made based on the total picture. There is little advantage to purchasing a specific computer and operating system and then finding that there are no appropriate software applications/packages available for that system. Mistakes can be costly both in terms of the actual purchases and in the personnel time and effort required to develop a functional system.

HARDWARE

Computers

There are primarily three 'sizes' of computers which can be considered for clinical trials applications:

- Multi-user mainframe computers
- Single or multi-user workstations
- Personal desktop computers (PCS)

It is unlikely that anyone would consider the purchase of-a multi-user mainframe system solely for the conduct of one or more clinical trials, and therefore this option is not discussed in detail in terms of selection criteria. If such a system is used for clinical trials, it is usually because it is already in use in a particular environment and the new clinical trial application can easily be implemented on the existing system. In such a situation, at minimum, the choices about hardware and operating system are already made. The choices about software may be limited to packages already available locally on that system. A multi-user mainframe system is usually well supported with operating staff and user assistance and should provide all the functionality that is needed for a trial.

Workstations are small but powerful computers that have replaced large mainframe systems in many computing environments. They can be single or multi-user systems and can usually be upgraded if more power/functionality is required. Most workstations have powerful graphics and mathematical processing capabilities and can be used as stand-alone systems or networked with other computers. In many environments workstations are set up in a network to share key disk space but provide the users with their own processing power. For example, the clinical trials database could be maintained on a disk that is accessible by all the workstation users, but when the users want to work on an analysis or report, they download the relevant files to their workstation and work on them there. This kind of environment is called client-server architecture, with each user having access to shared resources (e.g., the database) but with their own CPU (central processing unit) and memory. For large clinical trials applications where several people need to have access to the database for different purposes, this is a very viable choice in terms of hardware. As discussed later, many powerful database management and statistical software packages are available on these machines.

The microcomputer or personal computer is also a possible choice for clinical trials management. As these computers become more powerful, they provide much of the functionality that is required for a clinical trial. As single-user systems they are really only suited for small clinical trials where one person in handling all the data, but if networked together, do provide the capabilities needed for a multi-user, multi-trial system. There are clinical trials groups who use PC networks for some or all aspects of their operation. The limitations of this environment are usually the capability of the operating systems and appropriate software choices, but as time goes by, the choices are increasing and improving. It is important, however, that software be thoroughly researched to ensure that it has all the capabilities required. There are two main ranges of personal computers—the Macintosh range (Macs) and the "IBM compatible" range. The difference is in the architecture of the internals of the computer and in the operating system being used. Many manufacturers produce PCs that are compatible with the model first developed in the 1980s by IBM. There are a great variety of database and other software packages available for PCs, but many of them lack the power and functionality needed for large-scale operations.

Networks

While these three hardware possibilities are described independently, they can also be used together to provide the environment necessary for a trial. PCs can be linked to mainframes and workstations, and used with terminal emulation software, and machines of all types can be networked together. The two common types of networks are local area networks (LANS) and wide area networks (WANS). Local area networks are networks of computers in the same physical area that are linked together by cabling. There are different types of LAN operating systems, and expert advice should be sought before setting up a system. Wide area networks expand the network to link computers and devices that are not physically contiguous, and they can use telephone lines, microwave, or satellite connections to provide the links.

Peripheral Devices

The computer system alone is not going to be sufficient in terms of hardware. Peripheral devices such as printers also need to be available. Choosing a printer depends on the functionality required. Things to consider include speed, quality of output, volume of printing, and noise levels. Laser printers are now the most common choice for a multi-user environment, and these are available in many models that provide different levels of operation at a wide range of prices. If the printer is to be used by several people, will have a high volume of output, or if graphics are going to be produced, it is recommended that a printer at the mid to high end of the range be purchased. It is possible to have several printers on the same network, each providing different levels of functionality. With PCs and workstations, each user can also have their own dedicated printer. Ink-jet printers are viable options for this purpose and cost less than the laser printers. Color printers are becoming cheaper and can produce graphics and slides that are more interesting than ones in shades of black and white. Most networks will have a selection of these different types of printers accessible by all users, and choices will be dependent on the requirements and, of course, the available budget.

Other peripheral devices that may be desirable include fax machines and even copy machines. There are models of both that can be added to networks. Fax modems can also be installed in individual computers or made accessible through a network. These modems can be used for access via phone line to other computer systems as well as for sending and receiving faxes.

These are the most common types of peripherals that would be added to a clinical trials computing system. In sophisticated systems there may be other types of devices such as those allowing video-teleconferencing, on-line review of radiographic/photographic materials, or direct computerization of results of laboratory tests. Integration of all such devices should be part of the planning process, ensuring that all hardware is compatible and can be linked together to form an efficient and effective computing environment.

OPERATING SYSTEMS

The operating system of a computer is the software that governs the performance and capabilities of the actual computer. On some computers there is a choice of operating systems; on others there is only one available. If hardware has to be selected for a clinical trial, then it is important to take the operating system capabilities into consideration. Operating systems are machine specific, although there may be variations of the same operating system available on machines made by different manufacturers.

Mainframe operating systems are either proprietary systems developed by the hardware manufacturer specifically for that model, or variations of a widely used operating system, such as UNIX. An operating system developed for one mainframe machine cannot be installed on another make and model of computer, even if it is a commonly used operating system. This is because the construction of each computer is different, and the operating system controls the functioning of a specific design of machine. Mainframe systems have multi-user operating systems, allowing a variable number of users to access the same computer simultaneously. The end-user accesses the system by using a video terminal or a PC/workstation that can emulate a terminal interface.

Workstations are similar in that each manufacturer will develop one or more operating system(s) for their range of workstations. Manufacturers who produce a range of computer systems often develop a workstation version of an operating system that is also available on one of their mainframe systems. Versions of UNIX are common operating systems on workstations, although each manufacturer will have its own variation of the system.

On a personal computer, examples of possible operating systems are DOS, OS/2, Windows, and, on Macintosh computers, Mac OS. On IBM compatibles, Windows is an extended operating system that enhances the capabilities of DOS, the basic operating system. There can be different versions of an operating system. For example, the original versions of Windows, with version numbers up to 3.11, the newer Windows 95, and the network version, Windows NT. Most IBM compatible PCS come with versions of both DOS and Windows installed. DOS is a command-driven operating system, where the user has to type in commands and file names to perform specific tasks. Windows is a "point and click" application, where the user can use a mouse or other pointing device to select files, programs, and execute commands. Both provide the ability to use the computer at either a very basic level or a more sophisticated level, depending on the capability and experience of the user. The Macintosh also provides a "point and click" environment but does not provide the equivalent of the DOS environment. PC operating systems are typically for single-user environments, with the exception of Windows NT, which is a version of Windows that can be installed on a network.

There are other network operating systems available that allow management of multiple machines on a network. Most of them allow the individual machines on the network to be running an operating system that is different from the network management operating system and different from one machine to another. If both IBM and Mac PCs are to be networked, it is important to select a network operating system that allows both to be installed on the same network and to share the same network peripherals.

The key factors to consider when selecting an operating system are whether it should be a single- or multi-user system and, depending on the expertise and capability of the end-users and the scope of the application to be developed, features available and ease of use.

SOFTWARE

There are many types of software applications that can be used for clinical trials, but the key decisions to be made are about the software to be used for storage and maintenance of the trial database, and the software used for statistical analysis and reporting. The database software is discussed in detail in this chapter. Statistical software choices are usually made by the statistician for the trial and will depend on the analysis plans for the trial. Recommendations on specific statistical software packages are not made in this text because there are too many trial specific variables that need to be considered before making a choice. The focus of discussions here is on the interface between the database software and the statistical packages. Other types of software applications for clinical trials are discussed in Chapters 5, 6, and 10.

Database Management Software

The choice of database management system (DBMS) will depend on the projected size of the database and the functionality required. As described in Chapter 2, it is important that a detailed requirements analysis be done to determine these prior to making decisions on hardware and software. Possible constraints on choices will include cost and the available software packages for the specific computer(s) being used for the trial.

Most large or multicenter trials will require the use of a database with powerful data-handling features. Smaller databases could be managed using a spreadsheet program or the data-handling component of a statistical analysis package. However, for any environment where multiple studies will be conducted, or data will be collected from multiple locations, it is recommended that a DBMS be used. Many DBMS packages are available for different types and models of computer, and each will have a range of features. It is important to carefully review these features and to select the product which will most closely meet the requirements of the project. Table 4.1 lists some features that may be important in selecting a DBMS. In evaluating possible systems, first decide which of these features are important to the project, and then assess which are available in each of the systems reviewed.

In matching the list of requirements against the features available in a software package, there may be no single DBMS that meets all of the requirements. If this is so, there may be a temptation to try to develop an in-house DBMS and not use a commercially available package. This may appear to be the best option to provide the required functionality but is probably not realistic in the majority of situations. Development of any DBMS system from scratch is a very resource-intensive project and could limit future portability

Table 4.1. List of Database Management System Features.		
DBMS Features	Required for Project?	Available in DBMS under Review?
Database described using user-created data dictionaries		
Data entry screens easy to set up and maintain		
Support for direct data entry and, when applicable, data verification		
Ability to do ad hoc queries against the database without extensive programming skills		
Ability to have multi-user simultaneous read and write access to the database and maintain DB integrity		
Utilities available for data backup and recovery		
Adequate security features		
Interface with statistical analysis software		
Appropriate representation of missing values		
Support for a variety of data collection instruments and methods		
Documentation complete and easy to follow		
Training and technical support		
Appropriate user interfaces for technical and nontechnical staff		
Programming and report writing tools to simplify usage		
Cost effective		
Efficient storage and retrieval of the volume of data required for the trial		
Widely used by other sites, preferably ones involved in clinical trials		

options to different hardware and operating systems. This approach should only be considered in an environment where there are experienced personnel available, where there is time for the system development and testing, and where the system developed has the potential to be used for more than one project. A compromise approach would be to purchase the DBMS that most closely fits the project requirements and then to write applications software programs to customize the package for the project. It is important to allow adequate time in the development phase for the development and testing of the customized additions.

There are two database models that are most commonly used for clinical trials. The two models are the relational model and the hierarchical model. The following sections gives a brief description of these two models pointing out advantages and disadvantages for use in a clinical trials application.

RELATIONAL MODEL

The relational database model is currently the one most commonly used in commercial software packages. It stores data in tables, and each table is composed of columns and rows of data items. A single data item is called a field. A table is described by a data dictionary, and each row or record in a table has the same format. A database is made up of one or many tables, all related in some way. No two rows in a table can have the exactly the same value for all columns or fields.

Each table has one or more columns which are identified as key fields, allowing each row to be uniquely identified. This means that no two records in the table can have the same values in the key fields, although values in all other fields can be the same for all records. One table can be linked to another table in the same database by "joining" on identifiers that are unique between the tables. Figure 4.1 shows a simple example of a relational database structure. In this example there is a clinical trial database with two tables, one with baseline data for each patient entered on the trial and another with follow-up data on the same patients. Each patient has been assigned a unique patient ID for this trial and the patient ID is stored in a data field in both tables. The data records for one patient can be pulled from both tables by using the key field of patient_ID.

There can be more than one key field in a table, and the relationship between the tables can be one to one, one to many, or many to many depending on the number of records in the table for a particular patient. If there are multiple records per patient, then there does need to be at least one more key field in addition to the patient number; otherwise, the key is not unique. In this

BASELINE TABLE

Patient_ID	Date-on-study	Protocol	Weight
Patient_1			
Patient_2			

FOLLOW-UP TABLE

Patient_ID	Follow-up Date	Weight
Patient_1	Date_1	
Patient_1	Date_2	
Patient_2	Date_1	

Figure 4.1. Relational Structure.

example there is one baseline record per patient, and there can be multiple follow-ups. The key field in the Baseline Table is patient_ID, and in the Follow-up Table the keys are patient_ID and follow-up date. Patient_1 has one record in the Baseline Table and two records in the Follow-up Table, giving a one-to-many relationship, while patient_2 has only one record in each table and therefore has a one-to-one relationship.

Depending on the design and structure of the database, one table can be joined with multiple other tables to retrieve data. The key fields need not always be the same in each table. For example, if a third table (Fig. 4.2) called Protocol is added to the database in Figure 4.1, with two fields called "protocol" and "title," the records in the Baseline Table can be joined to the Protocol Table using the key field "protocol." This shows that data for multiple protocols could be stored in the Baseline Table and the Protocol Table if the records have the same structure.

There is no direct link between the Follow-up Table and the Protocol Table. To generate output that includes data from both, the join would have to be

PROTOCOL TABLE

Protocol	Title
protocol_1	title_1
protocol_2	title_2

Figure 4.2. Structure of Protocol Table.

done via the Baseline Table. For example, to generate information about the number of visits for a patient and include the protocol title in the output, there would be a join from the Protocol Table to the Baseline Table using "protocol" as the key field. The retrieval would find all records in the Baseline Table for that protocol and, using Patient_ID as the key field, would then join to the Follow-up Table and get all follow-up records for those patients.

This database structure is very flexible, although it can have high overhead in terms of processing, depending on the kinds of retrievals being done and the number of joins involved. With a very complex database structure, a large retrieval can take quite some time. An advantage of the relational model is the standard query language that has been developed. This language is called SQL (Structured Query Language) and has been implemented in nearly all major commercial relational DBMS packages. It is a fairly simple language that allows the user to query the database and retrieve data. It does require a knowledge of the table and data item names, and of the links or key fields between the tables, but is certainly a language that can be used by individuals without a programming background. Many of the database manufacturers are introducing user-interfaces that are very easy to use.

In clinical trials many tasks are oriented toward the data collection forms or screens. Data entry screens are created based on paper forms, data are collected using the forms, queries and edits are generated from the forms, and requests for overdue data are based on the presence or absence of the forms. The relational model allows creation of database tables (or views) that reflect the case report form structure and contents, and therefore it provides a close fit to functional requirements. The link to statistical analysis software can sometimes be complex because of this structure, but many of the commercial relational DBMS manufacturers have developed interface software for generating analysis files for the commonly used statistical software packages.

While not perfect, the relational database model appears to be one of the best choice for clinical trials applications. Most commercial packages will meet all or most of the criteria listed in Table 4.1. Ideally the package selected should have a fourth-generation programming language (4GL) allowing easy use by nonprogrammers. It is also an advantage if the package is available on multiple computer platforms (e.g., a PC model and a UNIX version). This allows for easier portability to different machines if that should become necessary.

HIERARCHICAL MODEL

This database model stores all data records in collections of hierarchies, with pointers linking one hierarchy to another. Figure 4.3 shows a representation

PATIENT HIERARCHY
Patient_1 Registration
Patient_1 Visit 1
Patient_1 Visit 1 Lab_1
Patient_1 Visit 1 Lab_2
Patient_1 Visit 2
Patient_1 Visit 2 Lab_1
Patient_1 Visit 2 Lab_2
Patient_1 Visit 2 Lab_3
Patient_2 Registration
Patient_2 Visit 1
Patient_2 Visit 1 Lab_1

PROTOCOL HIERARCHY
Protocol_1 Title
Protocol_1 Treatment_A
Protocol_1 Treatment_B
Protocol_2 Title
Protocol_2 Treatment_A
Protocol_2 Treatment_B

Figure 4.3. Hierarchical Model.

of two hierarchies—one showing patient related records and one showing protocol related records. For each patient in the Patient Hierarchy, there is a registration record, multiple visit records and, for each visit, multiple lab test records. The record contents must be structured so that each record can be correctly associated with other records for the same patient and can also be uniquely identified. For example, the visit records must be tied to the patient record, usually by a patient ID field. If the patient is on multiple protocols and all records are stored in the same database, then the registration records for the patient on a specific protocol must be unique. The visit records for each patient on a protocol have to have a field that will uniquely identify the visit which the record describes. This could be the date of visit or a sequential number assigned to the visit, as shown in the example. The lab results for the visits must also have a field that links them to that visit record.

The Protocol Hierarchy describes the protocols that are in the database and also the treatment arms for that protocol. Again, there needs to be a way to identify all treatment records for a particular protocol and this would normally be by means of a unique protocol number used in both records. Within each hierarchy we have therefore identified pointers or links between the records for a specific patient or protocol. The links are one to many (e.g., patient to visit) or many to one (e.g., treatment to protocol). The links become more complex when there is a need to link between hierarchies and introduce many to many relationships. Programming using this model is more complex than with the relational model, since the programmer needs to be very familiar with the database structure and navigate the database to produce the required output.

An advantage of the hierarchical model is that it usually more closely reflects the structure of a clinical trial and the data structure used for statistical analysis, although each statistical package still has its own required file structure. Therefore, there is still a need for programming support to produce the statistical files. There is also not usually a simple query language available for nonprogrammer users. In general, the relational model will be preferable to this one.

SUPPORT AND MAINTENANCE

All computer systems and environments require support and maintenance, and this should be considered when decisions about computing are being made. In an environment where a mainframe is in use, there is usually a computing support team already in place, and the end user can call on them with questions and problems. Much of the routine maintenance of the system such as backups, security, and software upgrades are more or less invisible to the user. The end user needs to be able to use the terminal (or PC) that connects to the mainframe, and also needs to know how to access and use the software and files relevant to the application. Often there are personnel available to help with software development and to help debug applications.

Some or all of these types of support will usually be available in an environment where computers are already in use for many purposes. When establishing a new computer environment for a clinical trial, it is important to consider how support will be provided. The main areas to consider are hardware support, software support, backups, security, and upgrades. All of these are likely to be required at some time during the life of the trial, and some of them, like backups and security, are relevant on a daily basis.

User Support

During the trial the Coordinating Center will be responsible for providing user support both to staff in the Coordinating Center and to trial participants at the sites. The responsibilities will include diagnosis of hardware and software problems and training in the use of any programs that are being used for the trial. Support for staff in the Coordinating Center is easier as the problem is on-site and the user support staff can actually see it and work to resolve it. Support for the participating sites is more difficult, since diagnosis and attempts to solve the problems will have to be done by telephone, often with inexperienced users trying to describe the problem and follow the instructions they are being given. This requires patience and understanding and, depending on the complexity of the trial, could take substantial personnel resources. Preparing detailed and easy-to-follow instructions for the sites will help to minimize the number of calls that are made for support, but the amount of time that will be required to answer these calls should not be underestimated.

Hardware and Software

Hardware is becoming more and more reliable, but the user can still encounter problems. Decisions have to be made about whether it is more cost-effective to cover the hardware with an annual or monthly maintenance policy or whether to accept that if something does go wrong, the full cost of repair and replacement parts will have to be paid. It is also important to keep track of software upgrades. Often internal file formats change when software is upgraded, and files need to be converted from the format used by the old version of the software to the format used by the new. The new version of the software will include the capability for automatic conversion of files, but if the user skips some upgrades and, for example, goes from version 2.0 to 4.0 directly without upgrading to the versions in between, the ability to convert from version 2.0 files to 4.0 files may not be included. Major upgrades to large database/statistical packages may require programming changes before the upgrade can be introduced.

Network Support

Maintenance of a network can be complex and time-consuming. As well as ensuring that all the computers on the network are functioning, peripheral devices such as printers, fax machines, and modems are usually part of the network. Maintenance will include the periodic upgrading of the network operating systems (as well as the operating system on the computers on the network), the backup of network files to ensure that files are not lost when there is a system crash, and the solving of hardware and software problems encountered by the users. In most network environments there is an individual designated for network maintenance, and this should not be overlooked when planning personnel for a trial environment. If a trial is heavily dependent on a network for computing applications, a lot of time can be lost if network problems are not quickly solved.

File Backup

An important consideration in setting up any computer system is deciding how backups of files will be done. Because of the possibility of hardware failure, it is essential that all computer files be backed up on a regular schedule. There are different ways of doing this. On a single PC the user can back up onto multiple diskettes, to a second internal or external hard drive, or to a cartridge or tape in an external unit. On a PC where a database is being maintained and there are frequent updates to the data, the backups should be done on a daily basis, and someone should be responsible for making sure that they are done. Doing backups is tedious, and it is tempting to think that the computer is reliable and there won't be a problem if it doesn't get backed up regularly. However, the consequences of losing several day's worth of data updates could be very serious, and it would probably take more time to recover from such a catastrophe than to routinely do the daily backups.

Network Server, Workstation, or Mainframe Backups

Backup devices can be added to a network server, a workstation, or a mainframe, and backups should be done on a regular basis, at least once a day and perhaps more frequently, depending on the volume of transactions. Some of the backups should be of the entire system, but others can be "incremental" backups of files that have changed since the previous backup.

A system needs to be developed for storage of the backup tapes or disks. Normally, backups are kept for at least a week, and one each week is kept for a longer time period. Some backups (perhaps monthly or quarterly) are kept indefinitely. The schedule should be developed before the trial starts and adhered to throughout the trial. As additional security, periodic backup copies should be stored off-site so that if there is a catastrophic disaster, not all copies of the database are destroyed. Periodic copies of a frozen database should also be archived, as should copies of the database and data files that were used for any major analysis of the trial.

Whatever schedule is used for backups, there is still a chance that data updated between the last backup and the crash will be lost. The database administrator needs to develop a system to ensure that transactions that were applied after the backup can be re-applied in the event of a crash. Most commercial database packages will provide software tools to ensure that recovery from crashes can be accomplished. While some data may have to be re-entered, following set procedures for backup and recovery should minimize the need for this.

Audit Trails

Most clinical trials generate a large database that is built and updated over a period of several years. New records are entered, and later, when more information comes in, corrections may be made to those records. It is important to keep track of the changes made over time so that there is a complete record of the transactions that constitute the final database that is analyzed. In smaller trials this "audit trail" could be on paper, where copies of all data are kept in the patient's paper record along with all correspondence, so that the record will always show how the database values were obtained. In larger studies the audit trail is usually an automated process and kept on-line. Most commercial database management systems have a journaling option that can help when the database administrator wants to cancel or roll back a transaction that failed or for recovering from a crash. Some systems will also provide a utility for maintaining a record of all database changes over time, as well as dates and the identity of the person who made the change. If this capability is required but not available in the DBMS being used, then either the paper audit trail must be used or software must be developed to provide this function. If an application is developed in-house for the trial, it should be thoroughly tested before implementation.

Security

It is essential that data collected for a clinical trial be kept secure and that patient confidentiality be maintained at all times. This applies to the paper records that are kept as well as any computer files for the trials. Paper records can be kept in locked rooms or file cabinets with access restricted to only those who are authorized. Electronic files should have the same protections in place.

Access to the computer where data are stored should be controlled by the use of log in IDs, passwords, and database security restrictions. Authorization to write to the database should be restricted, as should the ability to retrieve data from the database. Most DBMS systems have security features that allow you to restrict access to databases, tables, and even data items. In distributed systems there needs to be security at the participating sites as well as at the Coordinating Center. It is also important to ensure that investigators cannot access any data but their own unless such access is planned.

It is also important to protect as far as possible from computer hackers. There are different levels of security that can be introduced, and in setting up a clinical trials system, it is advisable to seek advice from someone who has experience in securing computer systems. If there is access via modems or the internet, password protection should be used. As with all passwords for computer systems, they should not be words that can be guessed easily and should be changed on a regular basis. A systems administrator should be responsible for checking for illegal access on a regular basis.

SUMMARY

Making decisions about hardware and software for a clinical trial is a critical part of the planning stage. Requirements and resources have to be carefully balanced before a choice is made, and it is important to ensure that the software being selected will provide the functionality needed for the trial. Peripheral devices should also be selected with care. The planning process should also include development of systems for backups, security, maintenance, and support. Without these resources the conduct of the trial could be jeopardized. Selection of the hardware and software is one of the most critical decisions to be made for a trial, since it will affect almost every aspect of the trial management.

CHAPTER 5

Data Entry and Distributed Computing

This chapter discusses some of the options to consider when selecting a system for entry of data into the computer. It first covers data entry done centrally in the Coordinating Center and then explores the possibility of distributing data entry and other aspects of computing for the trial to the local participating sites. This latter model has been successfully used in several large clinical trials but requires considerable resources and careful planning. Another option described is a hybrid model using facsimile transfer of data. As technology improves, all of these options are viable for a clinical trial, and the final decision should be based on a careful analysis of requirements and resources.

DATA ENTRY

For any clinical trial the transfer of data from the paper case report forms to the computer is a critical step, and accuracy of data entry is essential. Errors in data entry can obviously lead to wrong values being entered into the database and therefore can potentially cause erroneous results to be reported. It is extremely important that the data entry system be set up with adequate quality control checks. There are several ways of doing this including double data entry, manual review of keyed data, or computerized consistency checks of the data after it has been keyed—or a combination of the three. The method chosen will depend on the training of the person doing the data entry, the software that is being used, and the programming support available.

Data entry can be done either by professional data entry operators or by Data Coordinators or other staff who are familiar with the trial, and there are advantages and disadvantages in both approaches. Professional operators are fast and accurate and therefore can process large volumes of data in a timely way. However, they have no knowledge of the meaning of the data and therefore cannot correct errors or interpret poor handwriting on the forms. Records that fail the data entry error checks or where values cannot be read by the operator because

73

of poor handwriting will therefore be rejected in whole or in part depending on the conventions being used. The record will then require review by the Data Coordinator or another individual familiar with the trial so that they can try to interpret the value or seek clarification from the participating site if necessary. If data are keyed by the Data Coordinator, data entry will be slower and potentially less accurate unless there are additional automated checks on the data. However, the Data Coordinator will be able to understand and possibly correct errors as they are detected, depending on the type of error. When running small trials, this latter approach is usually acceptable, but for most large clinical trials the data entry is done by at least one professional data entry operator.

Software Selection

When choosing data entry software, it is important to look for a package that offers the features needed for the trial. Because the data entry is such an important step in the trial process, it is critical to have at least a minimal level of quality control done on the data during the data entry process. Using word processing software, text editors, or even spreadsheet software is not advised for data entry, since none give an adequate level of checking as the data are keyed. The word processor and editor do not even allow for the easy separation of data items into separate input fields of specific lengths. This can generate errors that are hard to detect afterward without checking all data carefully against the original forms. While spreadsheets have cells for individual data items, there are no checks on the values being entered and minimal checking of data types, and therefore using spreadsheets is also inadvisable. Purchase of a software package designed for data entry will be a wise investment.

Software packages designed for data entry provide some or all of the features that are listed in Table 5.1. Database management systems also provide data entry capability with varying degrees of sophistication. Table 5.1 lists some key features that should be considered when choosing data entry software. Not all will be applicable for every trial, and the final selection of software will depend on the requirements for that particular trial. The table also indicates whether the feature is typically available in the data entry tools found in most Database Management Systems, Data Entry packages, and word processors/screen editors/spreadsheets. This is intended only as a general guide. Any software package under consideration should be thoroughly assessed to be sure that it does in fact what is required.

Double- or Single-Data Entry

In many environments data are entered twice to ensure a high degree of accuracy. There is an ongoing debate about whether it is cost-effective to do dou-

Table 5.1. Common Data Entry Features.

Feature	DBMS	Data Entry Package	Word Processor, Editor, or Spreadsheet
Double data entry	Usually	Usually	No
Off-line data entry	No	Yes	Yes
Screens easy to set up and use	Usually	Usually	No
Range and field type checks	Yes	Yes	No

ble-data entry or whether an acceptable degree of accuracy can be achieved by other means. When double data entry is done, the data is keyed by one data entry operator and then keyed again, usually by a different person. If there are not two operators available then the data is re-keyed by the same person, usually with some time elapsed between the two sessions. There are two primary techniques available for double-data entry. The first creates two separate files with the data generated by the two operators, and the files are electronically compared. Discrepancies between the two are flagged, and the relevant case report form then needs to be checked to see which value is correct. Corrections can be made while the comparison program is running. In the second system, data are entered into a file by the first operator. The second operator then re-keys the data, and there is an immediate check that the value being entered is the same as the one that is in the file. Again, discrepancies are displayed and can be corrected after review of the forms. Double-data entry has been shown to have a high degree of accuracy with an error rate that can be as low as 0.001%. In large trials or trials that are being sponsored by industry and that may be filed with a regulatory agency with an application for approval of a new treatment, double-data entry is common.

The alternative to double-data entry is to enter data only once and then to introduce some supplementary quality control checks. The secondary checks can be by visual review of the data forms against the data entered or by developing computer checks of value ranges, field data types, and logical relationships between data items.

The **value ranges** check that the data item in a field is within an expected range of values. Sometimes this will be a single range, and sometimes the system will also check for a range plus other allowable values. For example, a data item could be acceptable if it is >0 and <25 or if it is 99, which is the code used for "unknown" for this field. If the system only allowed checking of one range, the range would have to be set to >0 and <99 to allow the "unknown" code to fall within range. This is not precise enough to make the range check meaningful, and therefore the ability to specify other allowable values outside the range can be important.

The **field type** check is one that verifies that the data entered in a field is of the correct type. For example, if a field is specified as "integer" and alphabetic characters are detected in that field, an error message will be generated. Many systems also allow date fields to be specifically defined, and any data in that field that cannot be interpreted as a valid date will cause an error.

Logical checks are usually trial specific and will require the development of a list of logical relationships between fields that can then be checked to ensure that the data "makes sense." For example, if gender is entered as "male" and a data item "Is the patient pregnant?" is answered as "yes," the program will detect the inconsistency and generate an error message. The extent of the logical checking will depend on the complexity of the data and for a large trial can be many pages of code.

Logical checks are usually run against the data after the data have been keyed, while range and type checks can be done as data are being entered. Since the logical checks are more complex, having them run in real time as the data values are keyed will slow down the data entry process, and therefore it is more efficient to run them all at once after the entry is complete. The logical checks can be extended to include checks between the new data and data that are already in the database for that patient.

Extensive checking of this kind can reduce data entry errors to an acceptable level and may be more cost-effective than having two data entry operators keying data. Another possible option is to do double-data entry only for identified key fields and to have the rest of the data keyed with single entry. If single entry is used, there should be reviews of accuracy of the key operator by introducing periodic double entry on some of the data. This is advisable even if extensive logical checks are in place. If data entry cannot easily be done at the Coordinating Center, there are contractors who provide this kind of service. If data is being sent off-site to a contractor for data entry, it is essential to ensure patient confidentiality. It is also advisable to develop an inventory checking procedure that keeps a record of all case report forms sent for data entry and makes sure that they are all returned.

On-line or Off-line Data Entry

Data entry can be done on-line (directly into the database) or off-line (into a file that is then updated into the database). The ability to enter data directly into the database is a tool that is usually found in most DBMS packages, and it therefore can eliminate the need for extensive setup of data entry screens. If a dictionary has been created for the record type, the range and type restrictions have already been defined and are automatically applied when data are being entered. The screen layout is usually automatically defined as well, although some DBMS will give alternative choices such as a full screen display of all or part of one record or a tabular layout where the data fields are displayed in one long record. This latter display usually allows entry of records from multiple forms without refreshing the screen, whereas the former displays only one record at a time. Other advantages of on-line data entry include rapid update of the database and fewer issues of security in dealing with intcrim files.

However, while the setup time for screens is minimized because of the use of the existing dictionary, there are potential drawbacks. Depending on the sophistication of thc DBMS, the screen layout that is created may not be conducive to data entry. To avoid confusion, the order of the variables on the screen must follow the layout of the form. Also, in the naming conventions for data items in the dictionary, the screen display should have meaningful names for the fields on the screen; otherwise, again data entry will be confusing. For example, if variable names in the dictionary are limited to a small number of characters and each has to be unique, there could be some names that bear no relation to the description of that field on the form. Other disadvantages are that overall performance of the computer system can be slowed down by direct data entry, and other users may be locked out of the database while transactions are being applied. Also, if the computer system is down for maintenance or repair, no data entry can be done.

Off-line data entry is done on a PC or other computer system that is separate from the database computer. The data entry software creates one or more files of all records keyed during one session, and the file is then transferred to the computer where the database resides. This means that if the database is not available, data entry can still be done. PC data entry packages offer a variety of features, and it is important to research them carefully so that they meet the needs of the trial. Because the PC system is unrelated to the database system, the screens have to be created independently of the database data dictionaries unless software can be written linking the two. While it is more time-consuming to create screens without the benefit of existing dictionaries, it does allow greater flexibility in defining the layout of the screen. The screen can be created to more closely reflect the layout of the form, and field labels can be made meaningful. All checks that are to be done also need to be defined and built in to the program. It is important to test all new data entry screens and associated checks for accuracy. There will often be modifications to the screens and checking routines as the trial progresses.

Using off-line data entry also requires development of software to transfer the data entry file to the database. This usually involves three steps. First, the keyed data are stored in a file format that is compatible with the database. Then the file is transferred from one computer to another. This can be done using diskettes (if the systems are compatible), a direct computer link, internet access or modems and phone lines, depending on the facilities available and the distance between the two machines. The last step is the update of the data file into the database. It is usually at this stage that extensive logical checks are applied to the data and error messages are generated. With off-line entry, systems have to be developed to ensure the security of the PC system and the integrity of the data transfer. Some data entry packages provide a password-controlled data entry system, allowing the data files to be accessed only by authorized users. Software should also be installed to guard against computer viruses. These programs scan the disks every time the machine is turned on and ensure that no virus programs are detected. To check the integrity and accuracy of the data transfer process, periodically files should be transferred twice and the resulting files compared. If there are discrepancies between the two, then there are problems with the transfer mechanism that need to be resolved. Until they are, it is advisable to continue with double transfer and careful checking or to use an alternative transfer method.

Screen Layout

Data entry screens are often designed to resemble the layout of the corresponding case report form. This allows the data entry operator to easily track the form and match the entry locations on the screen. This is one reason why case report forms should be planned with data entry in mind, since poor forms design can lead to data entry errors. (See Chapter 3.) Nevertheless, the screen cannot fully match the form for two reasons. First, the amount of text on a form is likely to be more than can fit on a single screen, and so meaningful abbreviations need to be developed for the field labels on the screen. Second, the standard computer screen allows only 24 lines of text, whereas a case report form often has more lines. Therefore either the software must allow the screen to scroll so that different sections appear as the data are reached, or there must be a facility to develop multiple screens for one form and to then link those screens together to create the database record. If a case report form has multiple pages, the pages also need to be linked. Large monitors that display a full page are available but are expensive. If the software does not allow linking of data from multiple screens, then software will need to be developed to either link the records after entry or to correctly update each record into the appropriate database table.

Colors and reverse video can be used to make data entry easier, but it is important not to incorporate too much detail on the screen or the visual effect will be confusing and the data entered difficult to detect. If any automated checking is done as data are keyed, then the program needs to display meaningful error messages on the screen. It is also normal for the program to beep to catch the attention of the operator and for further entry to be disallowed until the error message has been resolved.

Edit Checks

As mentioned above, there are various levels of automatic checks that can be done during data entry. The basic checks are range and type checks, and these features are found in all good DBMS and Data Entry packages. They can also detect missing values. Study-specific logical checks will need to be programmed, either using a programming language provided with the DBMS or data entry system, or by using a programming language such as C, Basic, or Fortran. The logical checks will be defined by review of the case report forms and specification of relationships between items and can be checks within a form or across multiple forms. If the latter type of check is done, it is done against the existing database for that patient after a new record has been keyed. Since logical checking can slow down both the rate of on-line data entry (if data entry is done directly into the database) and the performance of the system, the logical checks are better run after data entry is complete.

Conventions have to be developed to deal with missing values and with errors. If a field has a missing value with no answer filled in, the data entry software can be programmed to enter a standard code that indicates that the data were not available. If an error is detected during data entry or in logical checks run afterward, and the value is actually what is written on the form and not a transcription error, then a decision needs to be made whether the field is left with the value that caused the error or whether another code is entered to indicate that the value is being checked with the institution. If the value is found to be correct for that patient, even though it is an outlier and does not fall within the expected ranges for that field, it should remain in the database and flagged as "checked and correct" in some way. Different types of errors may be handled differently. In general, if any errors are detected in fields defined as "key" fields, then the entire record should be rejected pending review and correction of the data. If errors are found in other fields, the record can be processed and the field that caused the problem identified using one of the above methods.

The types of error checking done at the time of data entry will usually depend on who is keying the data. If data are being keyed by a professional data entry operator, the number of checks will be kept to a minimum, since the operator is not qualified to interpret the errors. This way the speed of data entry will be maintained. If data are being keyed by a Data Coordinator, more real-time checking can be done, since the Data Coordinator will have knowledge of the data and may be able to make ongoing corrections. Entry of data will, however, be considerably slower when extensive checks are being done as the data are being keyed.

It is very important that all data entered into the clinical trial database go through the quality control checks that have been established. No data should ever by-pass these checks regardless of when or how data entry is being done. This is particularly important when consistency checks are automatically done between fields. If one of the fields involved is updated outside of the data entry mechanism and the check is therefore not automatically run, inconsistencies can be introduced in the database.

DISTRIBUTED COMPUTING

When planning a multicenter trial, a choice has to be made about the method of data submission. At the end of a trial, there must be a complete database in the location where the analysis is done, usually the Coordinating Center, and therefore all data must be collected at the participating sites and Reference Centers and then transferred to this location. Most of the discussion so far in this text has followed the model where case report forms are completed at the participating sites and sent by mail to the Coordinating Center where they go through quality control checks and are entered into the computer. There are other possible ways of submitting data, and with the increasing availability of computers and software at a relatively low cost, electronic transfer becomes a viable option. The two most common methods of data submission are the paper model already described and a system where data are entered into computer files at each site and transferred electronically to the Coordinating Center. The Coordinating Center would be responsible for development and maintenance of the software required for data entry at the sites and for the electronic transfer. Copies of the software and subsequent upgrades are usually distributed electronically to the sites.

This second model has been used successfully in many clinical trials, and while it sounds appealing and appears easier than dealing with volumes of paper forms, there are several things to consider before selecting this method of data submission:

- What will be the volume of data at each site and centrally?
- How often will data be submitted?

- How many sites will be participating?
- Is the data submission system to be used for multiple trials or only one?
- How stable are the protocol and forms likely to be?
- What computing and personnel resources are available at each site?
- What resources are available at the Coordinating Center?
- Is there a need for the sites to have immediate access to their own data at all times during the trial?
- Is there a need for the Coordinating Center to have rapid access to the data collected?
- How many Reference Centers are there and what kind of data will be collected from these sites?

Volume and Frequency of Data Submission

The volume and frequency of data submission obviously have a direct impact on the number of personnel needed to do data entry. If volume is high and data are submitted often, then multiple people may be needed at each site to keep up with the data entry. Once this happens, the level of complexity increases, for there must be adequate personnel and computing equipment at each site to allow this. It is therefore important to adequately assess the volume of data that will be generated, not just the research data but also any administrative data that will be collected during the trial. Once the volume goes beyond that which can be handled by one person, either multiple single-user or networked PCs need to be available, or there needs to be a computer system that allows simultaneous access by multiple users. The ability of the field sites to maintain the system also needs to be considered.

Number of Sites and Trials

The more sites there are entering patients on a trial, the more complex a distributed system becomes. The Coordinating Center must be able to provide adequate training and user support for the participating sites on a continuous basis. There will be changes in personnel at the sites over time, and all new staff will need to be trained. Documentation will also need to be kept current at all the sites. If the system is to be used for multiple trials, this will also increase the maintenance costs, since all software programs and files for all trials will need to be kept current at the sites. If the sites are subject to frequent additions and removals, the distributed system may not be practical.

Stability of Protocol and Forms

While it is not optimal to make changes to a protocol or case report forms during the life of a study, sometimes it is unavoidable. Any such changes can obviously affect the data collection system; before any changes can be implemented, the revised protocol and forms will need to be distributed to all participants. With a distributed system this means that not only do paper documents need to be changed, but also electronic files and programs need to be modified, thoroughly tested, and installed at each site by the date that the changes take effect. If it is not possible to update all sites in one overnight transfer, both the old and new versions of the programs may need to be on the computers at the sites, and on the date of change, checks must be made to be sure that the new version of the software is being used.

The Coordinating Center must have adequate staff to make and implement the changes quickly. If the change is one that must be made before any new patients are entered on the study (e.g., one that involves patient safety), it may be necessary to suspend the study while the necessary programming modifications are made and the new routines tested and installed. In a paper-based system where forms are submitted to the Coordinating Center for entry, the programs running in the Coordinating Center will also need to be changed, but submitted forms can be held while the changes are made. Therefore, as long as the protocol and forms are modified and distributed to the sites, the trial can continue without suspension.

Resources at Sites and Coordinating Center

As discussed earlier in this chapter, data entry can be done by professional data entry operators or by trial staff. Professional operators will key large volumes of data quickly and accurately, while trial staff are not likely to be able to key with the same level of accuracy or speed. When data entry is done at the sites, professional operators would usually not be available unless the volume of data at the site is extremely high. The data are usually entered by CRAs, nurses, secretaries, or even physicians. Data entry will therefore be slower, but there is the opportunity for errors to be corrected in "real-time" as the values are keyed.

If there is a lot of complex data to be entered, it may not be practical for the entry to be done directly from the medical records; rather the data may be recorded on intermediate case report forms before any data entry is done. In this situation the workload at the site is increased by the introduction of distributed data entry, so adjustments will have to be made to responsibilities of the staff for there to be adequate time to do this additional work.

If data entry is done by field site trial staff, then extensive quality control checks should be built in to the data entry software so that errors can be detected and corrected at the time the data are entered. If new data need to be compared immediately with data already submitted, then a local database of all trial data needs to be maintained. Since there will also be a trial database at the Coordinating Center where additional corrections may have been made during the Coordinating Center quality control and update process, the issue of concurrency of databases needs to be addressed. If changes are made at the Coordinating Center, do they need to be transferred back to the database at the local site so that both databases have the same data? If a local database is not maintained, then the quality control checks against previously submitted data will need to be done by the Coordinating Center.

As well as ensuring that there are sufficient personnel resources at each site, the Coordinating Center needs to be sure that there is adequate computing hardware readily available for the trial. Ideally the trial budget will allow the purchase and installation of identical hardware and software at each site, thus simplifying the maintenance task for the Coordinating Center. However, if the trial depends on each center using or buying their own computers, the likelihood of problems increases, since a diverse set of computers is difficult to maintain from a distance. No matter how exhaustively software is tested on one PC, when it is moved to one that is "compatible," but not identical, there can be problems. There can also be difficulties with the availability of the hardware if there is not one computer dedicated to trial use. If the Coordinating Center has to develop software to run on different types of computers such as IBM PC compatible systems and Macintosh systems, the problems will be multiplied. Extensive maintenance of the systems at the local sites may be prohibitively expensive for the Coordinating Center, so it is important to have an adequate budget to cover this essential activity.

Access to Data

The need to access data immediately at either the Coordinating Center or the sites can lead to the choice of a distributed system, and there are situations where it would be necessary for the sites to have copies of all their data on a local database. For example, it would be important if patient entries were done using software that runs at each site and involves randomization techniques where treatments are assigned using dynamic balancing (i.e., when the new treatment is balanced based on exact information about previous entries at that site; see Chapter 6). Another example would be where the institution runs software to assist with patient scheduling. For such software to be accurate, there must be knowledge of the data already collected.

The Coordinating Center may also need rapid access to trial data. This could be for safety reasons, so that all adverse events can be monitored quickly or there could be a situation where subsequent decisions at the Coordinating Center are based on knowledge of data at the sites. For example, a treatment assignment (made by the Coordinating Center) may be dependent on results of tests done at the institution. The results of those tests would have to be available at the Coordinating Center before they could provide information about the subsequent treatment.

Distribution of Other Functions

It is possible to distribute other computing functions as well as data entry. The following are possible options:

1. *Distribution of the Entire Data-Processing Function.* This would include data entry, quality control of data, the ability to generate reports, and patient registration. Here the Coordinating Center is responsible for monitoring that the sites follow procedures. The Coordinating Center must also periodically collect all data and do the required analysis of data at the appropriate time points during the trial. If this method is chosen, rules need to be established before the trial activates about ownership of the data and the rights of the investigator to report on their own data independently of the rest of the trial. A trial could be jeopardized if results are discussed publicly before the study is unblinded. There also needs to be a decision about which database is the official one at specific time points during the trial. Once data are transferred to the Coordinating Center, there may be additional quality control done, and new variables may be calculated for analysis and stored in the database. At that time the databases cease to be identical, and the Coordinating Center must be the official one, such as will be used for the published reports of the study.
2. *Data Entry.* As described above, data entry can be done at the sites with on-line edit checks allowing for data corrections at the source. In this scenario, data are transferred to the Coordinating Center for all other functions and the official database is at the Coordinating Center. If additional quality control is done at the Coordinating Center, then queries can be transmitted electronically back to the sites for resolution.
3. *Patient Registrations and, where Relevant, Treatment Assignment.* If this function is distributed, the Coordinating Center will be responsible for installing appropriate software on computers at each site and then for monitoring that it is used correctly and that there is no abuse of the

system. If the sites are responsible for patient entry, a record of each transaction will need to be transferred immediately to the Coordinating Center where the total accrual to the trial is being monitored. Passing this responsibility onto the sites means that the sites are completely responsible for ensuring that the patients are eligible and that all regulatory requirements have been met. Usually the Coordinating Center will develop software to enforce adequate checking before the registration/randomization can be completed.

Central or Distributed System

In most instances the cost of implementing a distributed system is higher than that of a centralized system, and therefore the decision to distribute all or part of the computing for a trial can be made only after careful review of the requirements and the available resources for the trial in question. All of the issues described above should be considered before the decision is made. With a distributed system, it is important to recognize that the timing of implementation of all components of the system can be different between the central and the distributed systems. With the central system, the paper forms, protocol, and patient registration systems need to be in place before a trial is activated. If the data entry and database software are not completely ready, the incoming data can be submitted and held at the Coordinating Center until the programs are ready. However, with the distributed system, all software must be developed, tested, and installed at all the sites before the trial activates. Likewise, during the course of the trial, if there are changes to the forms or protocol, new versions of software must be installed at all sites before the changes can be introduced. This too can be a strain on the Coordinating Center.

OTHER DATA COLLECTION MODELS

There are other possible methods of data collection that can be used for a clinical trial. They are dependent on computer technology which is continually evolving.

Facsimile Transmission

There are several software packages available that allow collection of data at the sites on paper forms and then the transmission of the forms to the Coordinating Center by fax. Usually the forms have to be designed in a specialized format with designated fields for identification information, and forms development software is included as part of the system. Forms are completed at the sites and then transferred by fax to the Coordinating Center where they can be reviewed on-line. Digital images of the forms can be stored and retrieved and printed at any time, and data from the forms can be abstracted into records that are updated into the database. This method of data collection has been successfully used for several large trials. It is important to estimate an accurate volume of incoming data and to ensure that there are adequate phone lines and equipment at the Coordinating Center to handle that data. User support will also be important for this kind of system, both at the sites and the Coordinating Center. This system has the advantages of being able to print hard copy of paper forms if needed, but otherwise to have all data storage done electronically, removing the need for large numbers of file cabinets to store the paper forms.

Optical Scanning

Technology does exist to scan marks or data on forms. Either responses on a form are entered by making a mark in the appropriate area/box on a form, or answers are written using a character set that is predefined and can be interpreted by a scanner. Unfortunately, error rates for both technologies are still higher than acceptable for clinical trials data.

Direct Data Capture

More and more hospitals are developing internal systems that computerize all or part of their medical record. Laboratory test results are often stored directly in a computer system, and it is a tedious and error-prone process to transcribe these data onto paper forms. Any way in which already computerized data can be electronically transferred to the Coordinating Center will increase the accuracy of the trial data.

Technology also allows for storing digitized images of X-rays and other radiographic films. For example, it is feasible to have films transferred electronically from the treating institution to a Reference Center where they are reviewed. Results of the review can then be transferred to the Coordinating Center and become part of the trial database.

Most data for a trial are retrieved from the medical record for the patient. However, there are times when data can be captured directly into the computer. For example, instead of filling out a form, a patient can use a computer to complete a trial questionnaire, and the resulting data can then be transferred to the Coordinating Center.

All of these methods lead to capture of trial data at the source and can increase the accuracy of the trial database.

SUMMARY

The entry of data into the computer is a critical step for any trial, and any system that increases the accuracy of the process should be considered. Decisions need to be made about whether data entry is done centrally at the Coordinating Center or whether it is a function that is distributed to the participating sites. Software needs to be selected to meet the needs of the trial, whether data are being entered directly into the database or off-line, and then updated into the database. Checks on the data need to be defined and applied to all data entering the database. Whether data entry is done from submitted forms at the Coordinating Center or directly at the sites and transferred to the Coordinating Center to analysis, adequate quality control checks need to be in place. While data entered directly at the sites can be captured more quickly, careful consideration should be given before adopting a distributed system, since it is usually more costly to develop and more resource intensive to maintain.

CHAPTER 6

Patient Registration

Regardless of whether or not a trial involves randomization, it is recommended that patients be registered on to a trial *prior* to initiation of protocol treatment. As explained in Chapter 2, this can reduce any selection bias in patient entry and, because eligibility and regulatory compliance can be checked prospectively, will mean that the patients entered meet the requirements of the trial and will not need to be excluded from analysis for these reasons. Clearly, if treatments assignments are done by randomization, then prospective registration is absolutely necessary. There are several ways that registration/randomization can be done, and these are discussed in this chapter. There is also discussion of systems for suspending studies at times specified in the study design.

REGISTRATION

The term "registration" is used here to mean the process of entering a patient on a trial. Whether treatment is assigned or randomized, the registration process still has to be completed. Normally the process will require that the institution provide certain information about the patient being entered, and the Coordinating Center then provides the patient identifier for the trial and details about the assigned treatment. This process can be done using telephones, computers, fax machines, or using lists/envelopes of assignments distributed to each site. If randomization is involved, the system needs to have adequate security to ensure that investigators cannot influence or select the treatment for a specific patient or have any prior knowledge of the treatment that will be assigned.

The registration process should be structured and all patients should be entered following the same procedures. These could include:

1. *Verification That the Institution Entering the Patient Is Currently a Participant in Good Standing.* Institutions must be signed up to participate in the trial, have submitted any regulatory documents that are required, and not be suspended from participation for any reason.

2. *Verification That the Investigator Entering the Patient is Authorized.* For some trials, especially those using investigational treatments, investigators have to meet certain criteria before they can enter patients on the trial. For example, they may have to be approved by the sponsor or a national agency, or they may have to have certain qualifications. If there are requirements like this, then the registration process must include a check that the investigator is registered and approved for the trial.

3. *Verification That All Regulatory Requirements Have Been Met.* At minimum, most clinical trials require that each institution have a currently valid approval from an institutional Ethics Committee or Review Board. The registration process should check for this and also check that the patient has given informed consent according the laws and requirements of the country in which the trial is being done. Trials that are being sponsored by industry may also require the submission of other documents such as a list of laboratory normal ranges, investigator curriculum vitae, or a signed protocol document.

4. *Verification That the Patient Is Eligible.* It is important to be sure that the patients entered fit the eligible population as defined in the protocol. At minimum, the registration process should ask whether the patient is eligible, but in many trials there is an eligibility checklist that is completed during the registration process. Chapter 8 discusses eligibility checklists in more detail.

6. *Collection of Demographic Data.* At the time of registration, there may be a requirement to collect certain demographic data about the patient, either for monitoring purposes or as data needed for assignment of treatment. More details about this requirement are given later in the chapter.

The following is an example of a simple check sheet that can be completed at the Coordinating Center prior to registration to make sure that the above criteria are met:

EXAMPLE:

REGISTRATION CHECK SHEET
_____   Is the institution currrently active?
_____   Is the investigator authorized to participate?
_____   Date of most recent Ethics Committee Approval.
_____   Has the patient given consent?
_____   Is patient eligible?

This simple check can be expanded to meet the needs of the specific trial and can include confirmation of all eligibility criteria and collection of any necessary data.

Not all trials will require all of these steps. The requirements for a trial will depend on the sponsor, national regulations, whether investigational treatments are involved, and whether the trial involves assignment of treatment either through randomization or to a specific subset of patients. It is important that the necessary checks be in place and that the patients entered meet all the trial requirements. Entry of patients who do not meet eligibility criteria, or who are excluded from the analysis of the trial because of problems with regulatory details, can have harmful effects on the trial, and mean accrual goals may need to be increased to compensate. This can be very costly. The checking is relevant for any trial, whether the treatment is assigned directly or whether randomization is involved.

DIRECT TREATMENT ASSIGNMENT

In some trials either all patients or a specific subset of patients are assigned to a particular treatment. In a trial with a single treatment arm (Phase I or Phase II trials usually have one arm), all patients are assigned to that arm. The registration process will normally include an eligibility check, verification of regulatory requirements, and collection of any necessary demographic data. Once all these steps have been completed, a patient identifier will be assigned and the patient will be registered.

Some trials may have a more complex design where some patients have their treatment assigned by randomization and other patients with certain characteristics will have their treatment directly assigned. For example, in a breast cancer trial, high-risk patients (as defined in the protocol) may be randomized but low-risk patients all receive the same treatment. In such a trial questions that identify the low-risk from the high-risk patients must be asked during the registration so that the treatment can be assigned correctly.

RANDOMIZATION

Randomization is the term used when participants are assigned at random to one of two or more treatment arms that are part of the trial. Phase III trials almost always involve randomization. The technique is used when a promising new treatment is being compared to the current standard treatment regimen for that disease. Phase I and Phase II studies of the new treatment have usually been completed, and the treatment has shown some positive activity. The standard treatment arm is known as the control arm of the trial, and one or more new treatment regimens are compared with this control. Treatments are assigned using randomization algorithms developed by the statistician for the trial. The algorithms used for random assignment of treatment are discussed in the statistical literature and are not covered here.

Randomization is used to reduce any bias in the selection of treatment and to ensure balancing of patients across all treatment arms in a trial. Often the randomization involves balancing for several factors. Previous clinical trials in the disease may have identified prognostic factors—factors that have a significant impact on the outcome of treatment. For example, age or gender could be a prognostic factor. Clinical features of the patient's disease status could also be prognostic. For example, performance status or extent of disease. These factors are incorporated into the randomization design and ensure a balance of patients in each grouping on each treatment arm. This is known as *stratification,* and the prognostic factors that are used for balancing are called *stratification factors.* In multicenter trials there is usually also balancing by the participating site so that there is a balance of treatment assignments at each site and no site ends up with most or all of its patients on only one of the arms. Dynamic balancing is a term used to describe randomization systems that maintain complete balancing throughout the trial, taking into account any dropouts or unevaluable cases. A fairly sophisticated centralized computer program is required to implement complete dynamic balancing.

In randomized trials the data collected during the registration process will include any stratification factors that are being used for the trial. Since it is important to maintain an accurate record of the randomization process, the database should always accurately reflect the values given at the time of randomization. If a value is later found to be in error, the randomization record should not be changed. Instead, either a comment can be entered in the database or additional fields can be created to record the corrected value.

While prospective registration is strongly recommended for all trials, it is essential for randomized trials so that balancing can be achieved. Randomization can be carried out using one of several methods, depending on the resources available for the trial. The systems described here can be designed so that they can be used for nonrandomized trials as well as randomized trials. If a Coordinating Center is involved in many different trials, it is advisable to develop a registration system that can easily be adapted to suit different types of trials.

COMPUTERIZED REGISTRATION

A computerized registration system can be developed to run either at the Coordinating Center or at each participating site. If it is run centrally, the participants can either dial in directly to that computer system for the transaction or they can telephone the Coordinating Center and have one of the staff members run the program and work with them to complete the registration. A fully automated registration system for a randomized multicenter trial can be very complex, as it must do all the checking described above as well as assign treatments and patient identifiers. The following modules should be considered when developing such a system:

1. *Verification of Institution and Investigator.* As mentioned above, it is important to be sure that patients are being entered by authorized investigators at participating sites. In a fully automated system, there will need to be software to validate these items. This will require maintaining a database of all participants that can be accessed by the randomization software. The database will have to be current at all times and will usually have to allow for changes in status of both investigators and institutions over time. For example, institutions could be terminated or suspended as participants in the trial. Under either of these circumstances, registration should not be allowed.

2. *Verification of Regulatory Compliance.* If it is necessary to check Ethics Committee approvals, a database of most recent approval dates needs to be maintained. The randomization software will then need to check that there is a current date before proceeding with the transaction. Checking of consent can either be done by asking the institution to confirm that it has been done or by requiring submission of a copy of the consent prior to accepting the registration. This latter approach would be impractical in multicenter clinical trials where treatment needs to be started quickly if it caused unacceptable delays in registration. The treating investigator does bear responsibility for ensuring that consent is obtained.

3. *Eligibility Check.* At minimum, the participants need to be asked whether the patient is eligible, but more often all or a subset of the eligibility criteria for a trial will be checked during the registration process. If this process is automated, there need to be checks that the current version of the eligibility check is being used and that all questions are answered appropriately. Chapter 8 describes possible formats for checklists. The more detailed the checklist, the more complex programming that will need to be done. The program will need to allow for key entry errors and allow the user to correct wrong values. It may also need to allow for conditional branching where certain questions are asked only if the answer to a previous question has a particular value.

4. *Collection of Data Required at Time of Registration.* The software will need to have a mechanism for capturing the data required either for patient identification or as part of the randomization process (such as stratification factors). The captured data should become part of the trial database.

5. *Algorithm for Treatment Assignment.* The program must have a module that processes the data it has received and calculates the treatment assignment for the patient. This module will usually be based on a sta-

tistical algorithm for random treatment assignment and must be thoroughly tested and validated prior to activation of the trial. Any imbalance of treatment assignments can be costly as accrual may need to be increased to compensate. Therefore there should be periodic checks of the database to ensure that balancing is being achieved.

6. *Assignment of Patient Identifier.* As discussed in previous chapters, a unique identifier is usually assigned to each patient entered on a trial. This identifier would then be used in all subsequent communications about this patient, such as data submission or queries. The registration software should be responsible for assignment of this identifier and should keep a record of the number assigned.

7. *Notification of Treatment and Patient Identifiers Assigned.* Once the program has run, the participant needs to be notified about the treatment assignment and about the patient identifier. If the registration process is done completely by computer, the information needs to be clearly displayed on the user's video screen and should remain on the screen until the user issues a command to delete the screen. An option for the user to print the screen display will ensure that accurate information is recorded. As a backup it is recommended that the Coordinating Center generate a confirmation of the registration and send it to the participating site by fax, mail, or electronic transfer.

CONFIRMATION OF REGISTRATION

The following is an example of the kind of information that might be included in a confirmation of registration from:

CONFIRMATION OF REGISTRATION
TRIAL 0101

Date of registration:	6/16/98
Patient initials:	J.D.
Hospital	General Hospital
Hospital ID	123321
Age	>50
Gender	Male
Treatment Assigned	Arm A—Daily Aspirin for 2 years
Patient Trial Identifier	56789

The confirmation should contain sufficient information to verify the key aspects of the registration—the trial, institution, local patient ID, date of registration, assigned treatment, and assigned trial identifier. It documents the registration and ensures that there are no misunderstandings either at the Coordinating Center or at the site. If the site notices any errors when confirmation from the Coordinating Center is received, it must notify the Coordinating Center immediately.

In more sophisticated systems the computerized registration system may be linked to other processes for the clinical trial. For example, a registration could automatically trigger an order to a supplier to ship a drug for the patient who was just registered or a notification to a Reference Center to expect receipt of materials. The actual details will depend on the requirements of the trial, but the technology certainly exists now to be able to introduce these kinds of features.

CENTRAL OR LOCAL SYSTEM

As mentioned above, a computerized system can be one that is used directly by the participating sites or one where the sites call the Coordinating Center when they want to enter a patient and the staff at the Coordinating Center use the program to interact with the user and to enter answers given over the telephone. With this latter system the Coordinating Center may need to staff the randomization office for extended hours to allow for randomizations from different time zones. There will need to be sufficient staff to handle the randomization calls so that participants are not kept waiting for someone to call them back. If the trial is one where treatment has to start immediately and patients are critically ill, then it may be necessary to have staff available to take calls 24 hours a day, 7 days a week. If the Coordinating Center cannot ensure that there is always coverage, it may not be the best randomization system to implement for such a trial.

If the participants are using the randomization program themselves, they can either connect to the Coordinating Center computer system and run the program on that system or the Coordinating Center can install a version of the program at each participating site. If the participants use the Coordinating Center computer, they can connect using modems and phone lines or via the internet. Security needs to be assured when transmitting patient-related information using either method, and this needs to be factored into the system design.

If versions of the program are installed at each participating site, either the sites must upload data to the Coordinating Center or the Coordinating Center must have a way of polling each system to retrieve information about all the registrations that have taken place during a specified time frame. At minimum, this means every 24 hours. The software installed at each site needs to be programmed to balance entries on each arm at that site. When the data are reviewed on the central system, there needs to be software to check the overall balancing across all sites and, if there are imbalances, to make adjustments to the randomization software files at each site to bring things back into balance.

NONAUTOMATED CALL-IN SYSTEMS

If randomization software is not available for a trial, then alternative methods must be used. A paper-based system can be implemented to provide the functionality of the computer-based system described above. Participating sites will contact the Coordinating Center usually by telephone or fax, convey the necessary information, and the Coordinating Center will complete the registration using a paper-based system instead of a computer terminal. Such a system is more cumbersome to use and to maintain, but it will provide the necessary functionality. A paper-based system will depend on having paper versions of the modules described above. Many of the required lists will be generated by a computer system. The computer system will just not have the software available to complete an interactive registration/randomization. If no computer support is available at all, the paper files will have to be maintained manually. If this is the case, the person maintaining the system will have to be very organized and ensure that all manual updates to lists are entered appropriately. Keeping track of the system will require that strict procedures be maintained at all times.

1. *Verification of Institution and Investigator.* A roster of participating sites and corresponding investigators will need to be maintained and looked up at the time of a registration. Such a list would probably have to be created and updated using a word-processing program or database software, especially for large multicenter trials. If institutions or investigators are added or terminated frequently, then it may not be practical to edit the list every time and print a new one. If this is the case, then the printed list should have adequate space to write any interim corrections so that the registrar has current information at all times. If multiple people are involved in doing registrations, then all changes must be immediately conveyed to all of them.

2. *Verification of Regulatory Compliance.* If Ethics Committee approval or other regulatory documents need to be checked, then either those documents need to be readily available for review by the registrar in the Coordinating Center or a list of all current approvals needs to be on hand. The latter is preferable because it takes less time than going

through multiple files. If files are used, thought must be given to how those files are organized, especially if the Coordinating Center is responsible for several clinical trials and the registrars are involved with all of them. Logically the files can be ordered by participating site and, within the section for each participating site, ordered by protocol. However, many Ethics Committees will document approval of several protocols in one document, making it difficult to file by protocol. Also one Ethics Committee can be responsible for several participating sites. Therefore, if files are maintained, it may be necessary to make multiple copies of an approval so that all files are complete and relevant information available without having to track through a series of files. If a list of approvals is created, the list must also take these situations into account and have an entry for each protocol for each site.

3. *Eligibility Check.* A paper checklist can be completed at the time of registration with the Coordinating Center asking all relevant questions when the participant calls. If the registration is done by fax and not telephone, the participant would complete the checklist locally and fax it to the Coordinating Center where it would be checked before the registration was completed.

4. *Collection of Data Required.* Again, this information can be collected by telephone or fax and recorded on a form at the Coordinating Center. If stratification data are collected, there needs to be a method of accurately recording the values. It is also advisable for the registrar to repeat the values to the participant as a double check that they are accurate.

5. *Treatment Assignment.* In a paper-based system the computer will not be available to generate the correct treatment assignment, and therefore this has to be done manually. This can be done by generating lists at the time of study activation. These lists should be created by the Statistician for the study using software that has been thoroughly tested for accuracy. There will be separate lists for each institution and, for each institution, a separate list for each possible stratum. When a patient is entered, the registrar at the Coordinating Center will record the stratification data and then turn to the list for that institution/stratification combination. The patient would be assigned the next available treatment assignment on the list.

STRATIFICATION FACTORS

In studies where stratification is involved, there will be more paper lists for randomization because there needs to be a page for each possible combination of stratification factors. For example, consider a trial where there are three stratification factors with the following values:

Description	Value 1	Value 2	Value 3
Performance Status—Factor 1	0–1	2	
Age—Factor 2	<30	30–50	>50
Prior Treatment —Factor 3	Yes	No	

In this example, there are three stratification factors, two with two possible values and one with three possible values. The randomization system has to allow for any combination of these stratification factors. Using the notation 11 to represent factor 1 value 1 and 12 to represent factor 1 value 2 etc., the possible combinations of stratification values for this example can be shown as follows:

11 21 31
11 21 32
11 22 31
11 22 32
11 23 31
11 23 32
12 21 31
12 21 32
12 22 31
12 22 32
12 23 31
12 23 32

You can see that there are 12 possible combinations of stratification factors for this trial. The implications of having a large number of stratification values should be discussed with the statistician because, depending on the sample size, there may not be sufficient numbers of patients in each subset to allow any meaningful conclusions to be drawn at time of analysis.

For this example, there would need to be at minimum 12 assignment lists, one for each possible stratum. If there was also balancing by institution, there would need to be 12 for each institution. The following example shows a possible format for these treatment assignment lists:

EXAMPLE:

Institution: General Hospital Trial 1234		
Stratification 211 (Performance Status 2, Age <30, Prior Treatment = Yes		
Patient Initials	**Assigned Patient ID**	**Treatment Assigned**
A.B.	12001	A
G.J.	12005	B
		B
		A

In this example, two patients from General Hospital with Performance Status = 2, Age < 30, and Prior Treatment = Yes have already been entered on the trial. They were assigned Patient IDs 12001 and 12005 and treatment arms A and B, respectively. The next patient entered from this institution which fits these stratification criteria will be assigned to arm B. The assignment of ID can be done either using these lists and having the ID numbers preassigned (in which case they will not be assigned in sequence for the study) or separately using a "master" list of ID numbers for the trial, with a new patient being assigned the next one on the list. There would be another page for General Hospital for other combinations of the stratification values.

If this kind of system is used, the pages for each institution can be maintained in a loose-leaf binder with tabs for each relevant page so that the correct page can easily be located once an institution and stratum have been identified. Realistically only one book can be maintained, and therefore anyone involved in doing randomizations must look up that book and record the assignments and patient identifiers in that book as they happen. It is essential that the staff doing registrations understand that they should never convey any information about the "next" treatment to be assigned. Treatment information should only be given once all other registration steps have been completed.

MANUAL BACKUP FOR COMPUTER SYSTEM

The paper-based system described above can be used for all randomizations for a trial. It can also be used as a backup system for an automated computer randomization system. Because it is often essential that treatment be assigned quickly so that a patient can start on treatment, there must be some backup system in case of hardware/software failure. For large trials the backup could be on a second computer system, but more likely a paper-based system will be used. The system described here can be used for those purposes. A block of patient IDs can be preassigned for cases entered using manual backup. When the computer system is unavailable, the treatments are assigned from a backup listing. Usually, because small numbers of cases will be involved and the system will only be used (hopefully) infrequently, there is no balancing by institution in this situation. All necessary information is collected and recorded manually, and once the computer system is functional, all these data must be entered into the computerized system before any further randomizations are done. This ensures that the patients entered manually are factored in to any balancing being done by the software.

USE OF ENVELOPES AT SITES

An alternative to the paper-based system described above is to use a system where sealed envelopes are distributed to each site. Each site would have a set of envelopes for each stratum, and within the stratum each envelope would be numbered sequentially. The outside of the envelope would therefore have the name of the site, the stratum assignment, and a sequence number, as shown in this example:

Trial 0101

Institution: General Hospital

Stratum: PS 0–1, Age <30, No Prior Rx

Envelope Number 3

This envelope is for the third patient on the stratum from the site. Inside the envelope would be a registration form with a treatment assignment and patient identifier. As the patient is being entered on the trial at that site, an eligibility check is done at the site and the stratification values recorded. Once the overall stratum is determined, the next envelope in sequence for that stratum is opened. The treatment assignment in that envelope is the treatment assigned to that patient.Once the envelope is opened, the participating site can send information on the registration to the Coordinating Center so that a record of all registrations to the trial is kept up to date. The Coordinating Center in turn checks that the envelope opened is in fact the next in sequence. Enforcement of the system removes most chance for preselection of treatment at the sites. However, it is not eliminated completely. If there are two or more patients being entered from a site on the same day, it is possible for the investigator to open all the envelopes and, after reviewing the treatments assigned for all the cases, to select specific treatments for specific patients.

Balancing of treatment assignments is maintained within each site, but it is more difficult for the Coordinating Center to ensure balance across the entire study. It may be necessary to issue envelopes multiple times during the trial with overall balancing being done at the time of generation of each batch.

This system is mostly chosen when no computer system is available and when it is impractical to have all sites call the Coordinating Center for each registration, for example, when sites are spread across different countries and time zones. If envelopes are used, they should be thick and opaque so that the contents cannot be read without opening the envelope. It is important that all registrations be conveyed to the Coordinating Center as soon as possible.

REGISTRATION BY FAX

With the increasing availability and use of fax machines, it is feasible to consider using them for registration of patients on clinical trials. A fax-based system could be a sophisticated system where registration information is faxed to the Coordinating Center and captured as digital images by a computer system. The system interprets the data, assigns the treatment and patient ID, and faxes that back to the participating site with little or no human intervention. Such a system should require the transmission of all the information noted above and have all the necessary components for validation of that data. When necessary, data can either be reviewed on-line at the Coordinating Center or copies can be printed for review.

Fax machines can also be used in a more traditional way where the participating site faxes a form with all the needed eligibility and registration data. Staff at the Coordinating Center use that fax to complete the randomization and then manually fax the treatment assignment and patient ID back to the site. This latter model can be used if the Coordinating Center has either a computer-based or a paper-based registration system, and as long as the trial is one where patients do not need to start on treatment immediately, the fax system can be used in a trial where there are participants in different countries and time zones. The Coordinating Center staff would process any faxes as soon as they arrive at work in the morning so that treatment information is waiting for the site when work starts there. If treatment does need to begin immediately (e.g., if treatment has to be given to patients as soon after they have had a heart attack as possible), then the inherent delays in this system would not be acceptable.

USE OF TOUCH-TONE PHONES

Technology also allows the use of Touch-Tone phones and voice mail systems to register patients on trials. The participating site would call in and respond to prompts on a voice mail system. The voice mail system would be able to confirm certain values and then to give a treatment assignment and patient ID. While this system would allow registrations to be done 24 hours a day, 7 days a week, anyone who has used voice mail systems will realize that the registration process can be slow and have limited capability, and there certainly is no opportunity to ask questions or get clarification. If the value entered for each question has to be confirmed before going on to the next, the user could get frustrated. Also, if a trial had rapid enrollment and a complex randomization scheme, this type of system is unlikely to be practical.

BLINDED TREATMENT ASSIGNMENTS

All of the examples given so far have been ones where the treatment assignment has been "visible." Once the registration/randomization is complete, the Coordinating Center, participating site and patient are aware of the treatment being given. However, many randomized trials involve a comparison between one or more active treatments and a placebo. In these trials the information about the actual treatment assigned is kept from all but one or two necessary people. The Coordinating Center staff doing registrations, the treating physician, and the patient are "blinded" to the actual treatment. Nurses, clinical research associates, data managers, programmers, and statisticians are also blinded. A "double-blind" trial is one in which the physician (and other institution staff) and the patient are both blinded to the treatment; a "triple-blind" study is one where the statistician is also blinded. In reality, during ongoing trials the statisticians are usually also blinded to the actual treatment assignments. That is, although they may know the group of patients on each arm, they do not know the actual treatment assignment. In such a trial the active treatment(s) and the placebo have to be physically identical and administered in the same way. Blinded treatments can usually only be used when there are minimal side effects from the active treatment or placebo; otherwise, it would be obvious which arm the patient is on after treatment begins and the patient starts to experience the side effects.

A protocol with blinded treatment assignments should contain detailed information about when unblinding is to be done and by whom. If a patient is removed from the trial for any reason (excessive side effects, failure to respond to treatment, etc.), the treating physician will need to make decisions about any future treatment. If that decision depends on knowing the treatment that the patient received on trial, then the trial treatment will have to be unblinded. If the future treatment can be decided independent of the knowledge of the trial treatment, then the treatment does not need to be unblinded.

In situations where future treatment decisions need to be made quickly, there needs to be a mechanism for unblinding 24 hours a day. This can be done by telephone, with a contact being available round the clock. It can also be done by using special labels on the containers that hold the trial drug. Often the labels will have a tear-off feature where the pharmacist or physician can tear off part of the label to reveal the actual treatment assignment. Clearly there needs to be some protection to ensure that these labels are not tampered with while the patient is still receiving study medication and usually the participating site has to either retain all the labels for review at time of monitoring or send them to the Coordinating Center where they can be checked. Once a patient's treatment has been unblinded, they are usually considered to be off-study unless unblinding is part of the trial design.

When designing a randomization system for a blinded study, all these factors need to be taken into account and built in to the system. The randomization still needs to be done, but the actual treatment assignment is hidden to all but those who have an absolute need to know, such as those who prepare and ship the appropriate drug for the trial. Therefore the system has to assign the patient to the correct treatment arm based on all the same criteria as a nonblinded trial, but the information about the treatment must not be displayed on a terminal screen or on a confirmation of registration form. One way of doing this is to use the patient ID numbers as a means of randomizing the patient. Before the trial activates, a program would preassign treatment codes to each patient ID that will be assigned. The data generated by this program must be stored in the database (and on paper) in such a way that only a few authorized individuals can access the information. The randomization program will calculate the treatment that a patient should get and then cross-reference that treatment with the next sequence number that is assigned to that arm. The treatment description that appears to the user is usually displayed as "blinded treatment" with no indication as to the actual assignment.

The patient identifiers in such a system are not assigned in sequential order of ID number because of the randomization. Using this kind of system, labels for the medication used in the trial can be preprinted, and the medication can be appropriately labeled. Once the patient is entered and an ID assigned, the medication for that ID number is sent to the patient. The only people who have ready access to the actual treatment assignments are those who pack and label the medication.

If the trial is one where drug supplies are shipped to all participating sites prior to patient entry, then blocks of sequence numbers may also need to be allocated to each site. It is also possible to permit the pharmacist at the institution to know the actual treatment and to be responsible for labeling the correct drug for a patient. Sometimes this is necessary if the drug requires special preparation just before administration. If this kind of system is used, the pharmacist should be required to agree to maintaining confidentiality unless the treatment needs to be unblinded in a clinical emergency.

MONITORING ACCRUAL

The statistical design of a trial incorporates a calculation of the accrual required to answer the questions being asked in the trial objectives. Most trials have requirements for monitoring of results after a certain number of patients have been entered on the trial, and some require that accrual be suspended once that accrual level is reached pending completion of the review.

Phase I Trials

Phase I trials are designed to determine the MTD (maximum tolerated dose) for a new treatment and have a specified number of patients entered at different dose levels. The design requires that the specified number of patients be entered and their toxicity profile be reviewed before moving to the next treatment level. It is therefore essential in Phase I trials to have a mechanism for monitoring accrual and suspending the trial when the required number of patients has been entered on a particular dose level. For reasons of patient safety, this is absolutely critical, and the required accrual at a dose level should never be exceeded. It is also essential that accrual to the next dose level specified does not begin until after the required toxicity review has been done. Because this requires that the patients complete the treatment period and have their toxicity monitored and reviewed, there must be a mechanism for suspending accrual to the trial pending this review. It would be unethical to continue to enter patients while waiting for the review to be done. If registration is automated, then this ability needs to be incorporated into the software. If a paper-based system is used, the assignment lists can be clearly marked at accrual levels where suspension would be required. This is another situation where prospective registration of all patients should be mandatory.

Phase II Trials

Many Phase II trials have a two-stage design where a specified number of patients are entered and, depending on the number of responses to treatment that are seen, the trial may be terminated or continued. This is done to avoid accruing large numbers of patients to a trial whose early results indicate that the treatment is not as promising as hoped. With these trials there is often a need to suspend when the first stage of accrual is complete and to hold further accrual while the data on the first stage is evaluated. Of course, if the required number of events for continuing the trial have already been observed when the first stage accrual is complete, the trial does not need to be suspended. It can continue onto its second stage without interruption. If a trial accrued very rapidly, there is more likely to be suspension than for slow accruing studies.

Phase III Trials

Phase III trials are usually designed with early stopping rules built into the design. Statistical methodology exists to calculate the conditions under which the stopping rules apply. Under these rules, if a trial crosses the specified boundary (either positive or negative), it is terminated before reaching its final accrual goal. At these time points the data are reviewed by the Data Monitoring Committee for the trial, if one has been established. It is unlikely that a Phase III trial will be suspended pending review of one of these interim analyses, but it will possibly be suspended if there is an excessive rate of toxicity observed. Considerations of patient safety are always paramount in any clinical trial, and toxicity should always be closely monitored.

In any situation where a trial is suspended or terminated prematurely, there needs to be a mechanism for rapidly notifying all participants. This can be done by electronic mail or fax. Decisions need to be made about whether to still allow patients on study if they have already been checked for eligibility and given consent. This will usually depend on the reason for the suspension or termination. If it is for safety reasons, it may be unethical to allow new patients on study. It may also be unethical to continue to treat patients who have already been entered. These are issues that are considered by the Data Monitoring Committee.

For studies where accrual is rapid, it may be beneficial to establish a "waiting list" for accrual slots. With this kind of system, the participating site would need to call the Coordinating Center and reserve a slot prior to working up the patient and getting consent for the trial. In this way the Coordinating Center can notify the site if there are no further slots available and avoid the situation of getting the patient consent and then being told that the patient cannot go on the trial after all.

SUMMARY

Prior to activating a trial, it is essential to have a system in place for prospective patient registration. The system can be computer based or paper based and can be implemented either at the sites or at the Coordinating Center. The system must have the capability of implementing any preregistration checks that are needed and of notifying the sites about the treatment assignment and patient Identifier. When randomization is involved, the system is naturally more complex, and safeguards must be built-in to ensure the integrity of the randomization process and minimize any possibility of treatment selection. Blinded trials introduce a further level of complexity. Accrual monitoring should be part of the system with mechanisms to ensure that trials do not overaccrue. In randomized trials there should be periodic checks to ensure that treatment assignments remain balanced across all possible treatment arms.

CHAPTER 7

Local Data Management Systems

The role of the Clinical Research Associate at the participating site is a critical one for any clinical trial. The clinical staff who agree to participate in the trial are usually too busy to pay attention to the details that are part of the trial procedures and the CRA assumes responsibility for making sure that data are collected, tests are scheduled, and forms are completed and submitted in a timely way. The actual job description of the Clinical Research Associate will vary depending on the person's qualifications and the environment in which they work. Chapter 12 outlines a generic job description for a CRA, listing many of the possible responsibilities of a CRA who provides support for one or many clinical trials.

Organizational skills and ability to pay attention to detail are essential qualities for a CRA. Anyone in this position should also be able to work under pressure and be able to communicate with clinical staff, patients, and the staff at the Coordinating Center. It is a demanding role, and one that is becoming increasingly vital for the success of a clinical trial. This chapter discusses some of the procedures and tools that a CRA can develop to ensure that the trial requirements are met at the participating site. A CRA can contribute greatly to high performance and compliance in a clinical trial.

SELECTION OF PATIENTS

The faster a trial meets its accrual goals, the earlier the results of the research will be available in the medical literature. It is therefore important that all potentially eligible patients at a site be considered for entry on to the trial. Often in a busy clinical setting with multiple physicians involved in patient care, the possibility of entering a patient on a trial can be overlooked because the physician does not remember about the trial or is unsure of the eligibility requirements and does not have time to look them up. The CRA can assist the physician by screening patients and identifying those who might be eligible for the trial.

The screening can be done by reviewing a list of patients who have clinic appointments during the coming week or on the next day and, if it appears that a patient might be eligible for the trial, reviewing their medical record to get more details. A note for the physician can then be put in the medical records of potentially eligible patients, telling them that this patient may be eligible and listing what they need to do to confirm eligibility. If tests need to be done as part of the eligibility process, the CRA can be sure that they are scheduled. For each trial, a screening checklist could be prepared, and this could be the documentation that is placed in the medical record prior to the patient visit. Figure 7.1 shows a possible format for a simple screening checklist. This screening form could be expanded to include a list of tests that need to be scheduled to confirm eligibility. Any screening checklist will be trial specific and may need to be modified if the protocol changes during the accrual stage of the trial.

ELIGIBILITY CHECK AND PATIENT ENTRY

Once the initial screening has been done and the patient has agreed to be considered for the trial, it is important to go through all the eligibility criteria for

Trial 001, Eligibility Screening **This patient may be eligible for the trial of Aspirin/Placebo** **as a Preventive Treatment for Colon Cancer** **Please check the following During the Clinic Visit**			
	Already Verified	**To Be Verified at Visit**	**Outcome at Visit**
Within Age Range (20–60)	✓		
Performance Status <3	✓		
Family History of Duke's D Colon Cancer	✓		
Available for Long-Term Follow up		✓	
Willing to Participate		✓	
Not Allergic to Aspirin		✓	

Figure 7.1. Example of Simple Screening Checklist.

the trial and reconfirm that the patient is eligible. If any tests were scheduled as part of this process, it is important that the results of those tests be available for review before the patient is entered. The CRA should be responsible for this final check of eligibility regardless of whether the clinician says the patient is eligible or not. There may be an eligibility checklist included as part of the case report forms for the trial, and if so, it should be completed and then signed by the physician before the patient is entered. If there is no checklist included in the case report forms, then the CRA can develop a study specific checklist for use at that site. Suggested formats for eligibility checklists are described in Chapter 8, but whatever format is used, the checklist should cover all eligibility criteria for the trial and not just some selected criteria.

It is advisable to have all test results in writing when doing this final check and not rely on verbal reports. CRAs have found themselves in situations where the preliminary results given over the phone confirm a patient's eligibility, but when the written report comes in, the results are different and the patient is ineligible. This situation can be avoided by having the written report prior to finalizing the eligibility check. If necessary, request that the report be faxed so that there is no unnecessary delay.

If any of the eligibility criteria in the protocol are unclear, the CRA or the treating physician should call the Coordinating Center or Study Chair for clarification. However, this should not be used as a way to circumvent the criteria as written in the protocol. The patient should meet all criteria before being entered, and it is the responsibility of the CRA to confirm that this is true. If the treating physician puts the CRA in a difficult situation and insists that they enter a patient who does not meet all the eligibility criteria, it is important that the CRA give the *correct* information over the phone and not give erroneous information that would meet the eligibility criteria and allow the patient on to the protocol. Falsification of trial data is a serious charge and should be avoided at all costs. The CRA is therefore filling an important role in ensuring the integrity of the trial data, and if correct values are given to the Coordinating Center when entering a patient on trial, the Coordinating Center will detect the unacceptable values and not accept the patient. More details on the CRA's role in ensuring that standards of Good Clinical Practice are met are included in Chapter 9.

REGULATORY COMPLIANCE

The CRA should ensure that all regulatory requirements are met before entering a patient on trial and again can develop a checklist that is completed prior to entry. This checklist is a verification that everything has been checked and is in order. In the United States a checklist similar to that in Figure 7.2 can be

Trial 0101 Checklist for Regulatory Compliance Patient Initials: _____	
Regulatory Requirements	**Checked for Compliance**
Consent form signed and dated by patient and witness	✓
All regulatory documents on file at Coordinating Center	✓
Current Ethics Committee Approval for this protocol	Yes, approved 6/1/98 and date of entry is 10/1/98

Figure 7.2. Sample Regulatory Checklist.

used. In other countries the content of the checklist may change depending on local or national laws or regulations and the specific trial. If the trial involves investigational drugs or if a sponsor has additional requirements, the checklist can be expanded to include all relevant requirements.

PATIENT REGISTRATION

It is preferable for all registrations at a particular site to be done by the CRAs, since then they can be sure the patient is eligible and that all necessary regulatory documents are filed. The CRAs are also aware of all patients entered and can be sure that the correct procedures are followed. When a patient is entered on a trial, the local trial file should be initiated with copies of the checklists used for entry. Initial forms should be completed and submitted, and the schedule set up for future forms submission. If the CRAs are not notified of the registration (or do not do the registration themselves), some of these steps can be overlooked.

To register a patient, the CRA should follow the procedures defined in the protocol. If the registration is done by telephone, the CRA should confirm both the treatment assignment and the patient ID with the person at the Coordinating Center who is taking the call. Both should be written down immediately so that there is no room for error or misunderstanding. The trial folder for the patient should be created with a copy of the signed consent form (if applicable) and the registration information. Copies of correspondence and submitted forms can be added to this over time. If the medical records department will allow it, it is also useful to be able to mark the medical record in some way (e.g., a colored sticker on front) indicating that this patient is on a protocol. This means that anyone who accesses that record for any reason or treats the patient will know that there are potentially special procedures to be followed.

MAINTAINING FILES FOR THE TRIAL

The CRA will normally be responsible for maintaining the protocol, regulatory, and patient files for a clinical trial. Most hospital medical records departments have strict regulations about the types of documents that can be kept in the official medical record, and many of the documents related to a clinical trial cannot be kept in that file. Therefore, because it is important that copies of all trial-related data be kept, the CRA should keep copies of the trial data that cannot be filed in the official medical record in the trial record that has already been initiated. There is more information about this in Chapter 9.

The **protocol file** should contain a complete history of the protocol document over time including the following items:

- A copy of the original protocol and activation letter.
- A copy of all subsequent amendments and revisions to the protocol. If these amendments have to be integrated into the protocol document, it is recommended that there be at least one copy of each of the documents as circulated, as well as integrated copies of the revised protocol document. The former will provide a complete history of the protocol document, and the latter will contain the "current" protocol. It is important that both files be available for audits so that dates of changes to the protocol can be documented and it can be verified that patients received appropriate protocol therapy at the time it was given, even although the protocol may have changed since that time.
- Copies of all correspondence related to this protocol.

Copies of the *current* version of the protocol should also be maintained in appropriate places. This may include the doctors' office, the nursing station, the pharmacy, the clinic, and the ward. Any time that there is a revision to the protocol, the CRA should be responsible for ensuring that all copies of the protocol are updated accordingly.

The **regulatory file** for the trial will contain all copies of required regulatory documents including the following items:

- Copies of all Ethics Committee/IRB approvals for the trial.
- Copies of any signed contracts for the trial.
- Copies of the approved consent form for the trial, along with a history of any subsequent revisions.

it is also recommended that copies of all signed consent forms be maintained in a central location such as this file. This allows for easy verification that appropriate consent was obtained for all patients. Depending on the trial, there may be other requirements such as copies of laboratory certifications and documentation of ranges of normal values for laboratory tests, curriculum vitae of investigators, and other regulatory documents required by the sponsor or the local government.

The **patient files** for the trial should also be maintained by the CRA and should contain copies of all forms submitted to the Coordinating Center, copies of all queries received from the Coordinating Center, and copies of all responses to those queries. Any patient-specific correspondence should also be kept in this file. Another important addition to this file should be copies of any data relevant to the trial that might be stored at other hospitals or doctors offices. For example, any interim laboratory work done on patients between treatments, such as tests done by their local doctors rather than by the trial specialists. It is important that copies of these data be obtained on an ongoing basis and stored as part of the patient's trial record. These are data that will be needed if the case is audited, and it is much more efficient to routinely request the data from the source rather than having to do it all at once.

It is recommended that each patient's records be filed in a separate folder and that the folders be filed according to patient identifier rather than patient name. The Coordinating Center for a trial usually does not have access to the patient's name, and therefore all correspondence and communication from the Coordinating Center will use the patient ID number. If the files are organized according to this number, it will make it easier to deal with the communications. It is also recommended that the CRA maintain a cross-reference file of names and IDs so that the patient can be identified regardless of which identification information is given.

PROTOCOL COMPLIANCE

While it is the responsibility of the clinician to ensure that the protocol treatment plan is followed, the CRA can provide support to the physician to ensure that this is done. Some CRAs include a copy of the entire protocol document or selected sections of the protocol in a patient's chart. This is useful because it means that the treating physician always has a copy of the protocol on hand when seeing the patient. However, the protocol can be bulky, and the system can be problematic when there are amendments to the protocol and all the files have to be updated. An alternative method of ensuring compliance is for the CRA to prepare a summary of relevant information for the clinician prior to a patient's visit.

Before a patient's visit, the CRA reviews the patient's record and makes a list of tests to be scheduled and data to be collected during that visit. This information can then be put inside the patient's record so that the physician is reminded of the protocol requirements when looking at the chart. The more specific the "instructions," the more likely it is that the relevant data will be collected and recorded. If tests are to be scheduled either during or after a visit, the CRA can develop specific tools that help to make sure that the necessary appointments are made. Because most protocols require all patients to follow the same procedures on a fixed schedule, the tools can be developed ahead of time and used when appropriate for a patient. Figure 7.3 shows an example of an internal form that has been developed to ensure collection of toxicity data and scheduling of the required laboratory tests for a protocol.

Trial 0101
Visit – Month 3
Patient ID _____   **Patient Name** _____

Please be sure that the following information is collected from the patient during this visit

Questions for the Patient

 1. Information about any side effects that have occurred since the last visit.
 2. Has the patient taken the medication according to the protocol schedule?
 3. If there were any interruptions in taking the medication, why did they happen?

Please Do the Following Test at This Visit

 1. Physical exam, including BP, Weight, and Performance Status

Please Schedule/Arrange the Following during the Visit

 1. Draw blood for CBC
 2. Chest X ray

Figure 7.3. Example of a Worksheet for a Clinic Visit.

This kind of format can be modified to suit the requirements of a particular trial. It can also be modified so that the physician can actually record the data on the form. Any data written on the form should be signed/initialed by the physician and kept in the patient's file.

Any tools that the CRA develops and uses to improve protocol compliance can only improve the quality of the trial data.

PATIENT'S ROLE

It is usually a good idea to get the patient involved in the details of the trial, for this will help in ensuring trial compliance. The more patients know about the tests to be done and the schedules for visits, the more likely they are to comply—and to remind the physician or CRA if things are not done. Patient calendars can be developed and given to each patient when they enter the trial. The calendar could contain information about the number of visits and the schedule of visits. If the date of registration is used as the starting point, the actual dates of visits can be filled in at the start of the trial and the appointments made. For each visit there can be a list of the tests that will be done and information about the treatment to be given during that visit (if relevant). This kind of approach makes the patient more committed to the trial and therefore more likely to complete the requirements of the trial.

If the trial involves the patient taking medication at home, then a patient diary can be developed where the patient can record details of the times and doses of treatment given. This then becomes a record of the patient's compliance with self-medication. In some trials this diary is part of the records to be kept for the trial and is developed by the Coordinating Center. However, in trials where there is no official patient diary but medication is being taken at home, it is a good idea to create one and, at the end of the patient's treatment, to collect it and to store it in the patient's file. It can then be available during audits and can be verification that the patient followed the trial. The patient can also use the diary to record any side effects of treatment as they happen rather than trying to remember them at the time of the next clinic visit. This helps to gather more complete side-effect data for the trial.

COMPLETION OF CASE REPORT FORMS

Clearly one of the most important responsibilities of the CRA is to complete the data collection forms (paper or electronic) that are required for the trial and to ensure that they are sent to the Coordinating Center in a timely way.

When a new protocol is activated, the CRA should make sure that they have an adequate supply of blank forms. If the forms are on NCR (no carbon required) paper or in specially prepared booklets, then copies should be "ordered" from the Coordinating Center, and as supplies are used, replacements should be requested. If the forms are part of the protocol document, a master copy of the blank forms should be filed so that additional copies can be made as necessary.

The CRA should familiarize themselves with the data submission requirements for the trial and with the contents of each form. If the data collection process will involve the input of people from other departments or units, it is advisable to have a "training meeting" for all people who will be involved in the data collection so that they all understand their responsibilities for the trial. If questions arise when the forms are being filled out, the CRA or physician should call the Coordinating Center for answers. It is better to take the time to get clarification on something that is unclear than to have to deal with queries from the Coordinating Center because the forms were not correctly filled out. Asking questions also lets the Coordinating Center know that there are potential problems either with the forms or with the instructions for completing those forms.

FORMS SCHEDULING

Once a patient has been entered, the forms calendar for that patient can be initiated. This is a patient-specific projection of the dates that forms will be due for the patient if they follow the protocol exactly. Figure 7.4 shows an example of a possible format for a patient calendar. In this example the trial involves an initial visit followed by 3 monthly visits and then a final visit after 8 months. The dates for the forms can be projected based on the patient's registration date and will provide a reminder about forms for this patient. The calendar may need to be modified over time if there are delays in visits or if certain events intervene in the schedule. For example, if the patient goes off treatment at visit 2 because of severe side effects, the forms for visit 3 will not be required.

This simple calendar system will work if a CRA is responsible for a few patients on a single trial. However, if the CRAs are following many patients on several trials, they need to have a better system for reminding them when forms are due for all the patients. As mentioned in Chapter 2, such a system can be computerized if the resources are available to develop the necessary software and files. A computerized scheduling system could have the following features:

TRIAL 0101
PATIENT CALENDAR

Patient ID _____   **Patient Name** _____

Date Entered: 6/1/98

	Date Due
Initial Visit	
Demographic and History Form	_____
Baseline Test Form	_____
Follow-up Visit 1	
Side Effects and Complications Form	_____
Medications Compliance Form	_____
Follow-up Visit 2	
Side Effects and Complications Form	_____
Medications Compliance Form	_____
Response Assessment Form	_____
Follow-up Visit 3	
Side Effects and Complications Form	_____
Medications Compliance Form	_____
Final Assessment Form	_____
Final Visit	
Long-Term Effects Form	_____
Disease Status Form	_____

Figure 7.4. Patient Forms Submission Calendar.

- *Ability to Handle Multiple Protocols.* The system has to include a definition of the forms and their submission schedule for each protocol that is part of the system.

- *Ability to Handle Multiple Patients.* The system should allow entry of relevant data on multiple patients and be able to generate forms submission schedule reminders based on the relevant data for a particular patient and protocol.

- *Ability to Adjust Schedules as Events Occur.* While some protocol forms are due according to a fixed schedule, others may be due when specific events occur. For example, if an Adverse Event occurs, a specific form usually has to be submitted immediately. The program therefore needs to be able recognize that this form is needed and generate a reminder.

- *Ability to terminate further data submission reminders when a patient is no longer on protocol.* This may happen according to a particular schedule (the protocol treatment plan) or because of an event such as a patient's refusal to continue on protocol or failure to respond to treatment.

To ensure that the system is flexible and can handle changes in the patient's status or specific events, the CRA will have to continually update the database to add relevant new information about the patient. If the patient-specific data are not kept up to date, then the more advanced features of a scheduling system will not be able to generate accurate information. It is strongly recommended that any such scheduling system be "data driven." This means that there would be data files accessible on-line by the program containing information about the "rules" for a specific trial. With this kind of design, when a new study is activated, or a change is made to an ongoing trial, these data files need to be created or updated, but the program itself does not need to be changed. This reduces the resources required to maintain the system, and after the initial programs are written and tested, the maintenance can be done by an experienced CRA rather than by a computer programmer.

If programming resources are not available, the CRA can develop a paper-based system to help keep track of forms submission requirements. While this may perhaps be not quite as appealing as a computer system, it can be just as effective. The system could be implemented by using "cards," or a desk/wall calendar. The latter is most effective when there are only a few patients to follow. If there are larger numbers of patients on multiple trials, the card system is probably a better choice.

Calendar System

When a calendar is used for keeping track of patients, the CRA enters information about a patient under the month/week/day when data are due. There should be sufficient information to be able to easily identify the patient and the form that is due. This would usually include at minimum the trial number, patient ID, and identifiers/titles for the forms that are to be completed at that time. All projected requirements for a patient could be filled in when the patient enters, or the data for the "next" due date could be entered when the "current" forms have been completed. As forms are sent, the entry on the calendar should be checked off or crossed out so that the CRA knows the forms have been completed and sent.

With this system only a small amount of information per patient can be maintained on the calendar because of limited space for writing. Additional

---

March 1996

Trial 0101—pt 15432—Visit 3 forms Due
Trial 0101—pt 15321—Final Visit forms

Trial 0201—pt 12123—Submit chest X ray

Remember to get Ethics Committee reapproval of 0101

Figure 7.5. A Calendar System Entry.

information (e.g., the date that forms are actually completed and sent to the Coordinating Center) could be kept as a reference in the patient's trial record. Figure 7.5 shows an example of how entries could be written for a particular month.

Card System

When using the card system, a 3×5 inch (or similar) card file is created with a section for each week or each month, depending on volume and submission requirements. When a patient is entered on the trial, a card is created showing what forms are due and projected dates. This would be similar to the information shown in Figure 7.4. Once the initial forms for a patient are completed and submitted, the card would be filed in the section of the file for the month/week when the next form is due. For example, for the patient entered in June 1996, the initial forms would be completed and the card would be filed in the section for July 1, 1996. Each week/month the CRA would pull the cards for that time period and make sure that all the relevant forms are completed and submitted on schedule. The information on the card could be expanded to include a space to record the date of submission to the Coordinating Center.

While this kind of system is clearly not as sophisticated as the computerized scheduling system, it can still be effective in environments where there are no resources to develop a computer system. It can be enhanced by data requests and reminders sent out by the Coordinating Center.

It is important to know that forms have been submitted to the Coordinating Center and have not been forgotten. This can be done by recording the date sent, as noted above and/or by writing/stamping the "date sent" on the forms itself. This allows you and the Coordinating Center to keep track of data that may be in the mail and will also identify any unacceptable delays in the transmission of data to the Center.

SUBMISSION OF MATERIALS

Often a trial will require the submission of materials other than case report forms. Examples of such materials are X rays or scans, pathology materials, blood/tissue samples, or photographs. When a trial has such requirements, it is the CRAs responsibility to be sure that they are available and submitted as specified. These requirements can be added to the "reminder" tools that have already been discussed.

MONITORING TRIAL INVENTORIES

There may be special supplies needed for a specific trial. For example, copies of forms packets, shipping containers for samples for reference centers, and a stock of the medication/devices being used in the trial. The CRA should develop an inventory system to ensure that there are always adequate supplies available for current and future patients. These can be simple lists of supplies received, amount used, and balance, but they should be checked regularly and restocked before the are urgently needed.

DOCUMENTATION OF PROCEDURES

While the Coordinating Center will be responsible for preparing and circulating documentation of procedures for a clinical trial, it is also important that the CRA maintain documentation of systems that are specific to that location. This could include the name and phone number of contacts in other departments in the hospital who can be relied on to help gather necessary information for a trial, information about the Ethics Committee requirements on format for submissions, and the average time lag in getting protocols approved or reapproved or details of hours that the pharmacy is open. It is hard to give specific examples here, since the issues will be specific to the particular location and trials being done. If you document everything that you think a new CRA would need to know to be able to work on the trial, you will have a worthwhile and important document!

SUMMARY

While the clinician at a participating site is ultimately responsible for the conduct of the trial at that site, the CRA acts as an agent of the clinician and is responsible for implementing most of the detailed requirements of the trial. The CRA role is a vital one and, if done properly, will lead to the collection of complete, high-quality data for a trials. While it takes time to set up the procedures that are described here, having them in place and working effectively can lead to increased efficiency during the trial. Audits are greatly simplified if all data are accurately recorded and filed in the appropriate place. Extensive organization and the ability to pay attention to all the details will help ensure that all trial requirements are met. The CRA is an essential member of the trial team at the participating site and can be instrumental in the successful conduct of the trial at that site.

CHAPTER 8

Central Quality Control of Data

All clinical trials data should undergo quality control checks to ensure that they are as complete and accurate as possible. For multicenter clinical trials it is the responsibility of the Coordinating Center to implement quality control checks on all data that are submitted from participating institutions. The Coordinating Center has no access to source documentation for the data and must work only with the forms and other materials collected for the trial. It is therefore important to ensure that submitted case report forms have been completely filled out and that data are consistent. While the procedures described here best fit the model of a multicenter clinical trial with a Coordinating Center, it is equally important to do quality control on data collected for small, single institution trials, and it is recommended that the quality control process involve review of the data by someone other than those directly involved with the treatment of patients entered on the trial and the initial collection of the data. Any quality control process should provide a consistent and objective review of data and assessment of study endpoints. The checks can be done manually, electronically, or a combination of the two.

To ensure that checking is done consistently across all patients over time, a manual of procedures should be developed for a trial, documenting the checks to be done on all cases, and the action that should be taken if errors are detected. Inconsistencies will normally require a query back to the participating institution, but some may be able to be resolved centrally. For example, inconsistencies introduced because of a data entry error do not need a query to the institution. Once the error is corrected, the data will pass the data entry check. This chapter outlines the kinds of checks that should be considered and suggests possible ways of implementing them.

ELIGIBILITY CHECKING

Every protocol should contain a section that describes the patient population to be studied in the trial and lists characteristics/criteria to be met before a patient can be entered. It is very important that the protocol is always followed and that the eligibility requirements are met by all patients entered. Entering ineligible patients on to a trial can compromise the validity of the results, and if an excessive number of ineligible patients are entered, the sample size may have to be increased. At best, the increase will be equal to the number of ineligible cases entered, but if the planned analysis for a trial will include all patients (an "intent to treat analysis"), the ineligible cases can dilute the analysis; to compensate, the accrual goal may need to be increased by a number that is higher than the number of ineligible cases entered. As patient resources for trials are limited, it is important to use them wisely. Ineligible entries serve no useful purpose, and there is no advantage to making exceptions to the criteria defined in the protocol to enter patients who "almost" meet the eligibility criteria, but not quite. At the local institutions, eligibility checks can be done prior to registration to make sure that patients meet the criteria. It is recommended that the person entering the patient on the trial have hard copies of all data needed to confirm eligibility and that checks do not rely on information given verbally; for example, an actual written pathology report should be present rather than using information given by telephone.

At the Coordinating Center there should be a system in place to ensure that patients are indeed eligible before they are accepted on to the trial. The checking can be done in several ways. It is of course possible to assume that the personnel at the local site have checked the patient's eligibility and to accept the registration if they say that the patient is eligible for the trial. However, it is more common now to have an eligibility checklist that must be completed prior to entry on the trial. This checklist becomes one of the case report forms for the trial. It can be completed by the local institution prior to registering the patient, and a copy of the checklist sent in as part of the data submission requirements. Alternatively, if the patients are registered by a call to the Coordinating Center or by entering data directly into a computer system, the responses can be stored as part of the record of the registration.

An eligibility checklist has a list of questions that must be answered to determine that the patient meets all necessary criteria. When designing an eligibility checklist, it is important to phrase questions in an unambiguous way and not to construct awkward questions to force a specific answer. For example, if the checklist requires that all questions be answered "yes," the following question could cause confusion: Is the patient not pregnant? The question would be better worded as: Is the patient pregnant? with "no" as an allowable answer. Trying to oversimplify questions on any form may lead to confusion and should be avoided. The questions should be posed in the most natural manner and be clear and unambiguous, especially in circumstances where the first language of the person filling out the form is not that in which the questions are asked.

A simple format for eligibility checklists is one with a section of questions where the answer must be "yes" and a section where the answer must be "no." Any question where the answer is "unknown" would make the patient ineligible unless the answer can be obtained. Because the format allows for both "yes" and "no" answers, the questions are clear and unambiguous. The checklist can be completed quickly and easily prior to entering a patient on a trial. Example 1 shows a checklist prepared using this format.

EXAMPLE 1: A simple checklist where "yes" and "no" answers are allowed and where the questions are grouped according to the allowable answer.

Eligibility Checklist		
Protocol: _____ Patient ID: _____ Institution: _____		
Yes No		
__ __ Is patient 15 years of age or older?		
__ __ Is there a confirmed histological diagnosis of colon cancer?		
__ __ Is there evidence of metastatic disease?		
__ __ Is the date of surgery more than 21 days and less than 60 days from today's date?		
__ __ Were all required laboratory tests done within the last 14 days?		
__ __ Is the patient's WBC $>4000mm^3$?		
Note: The answers to all questions in this section must be "Yes"		
Yes No		
__ __ Is the patient pregnant?		
__ __ Has the patient had prior treatment for this cancer?		
__ __ Has the patient any history of cardiac problems?		
Note: The answer to all questions in this section must be "No"		
Patient is eligible for the protocol if all above criteria are met		

This checklist is simple to design and easy to complete. However, because the answers are predictable, there could be a tendency to rush through the completion without fully checking that the answers are correct and verifiable. When pressure is put on a CRA by an investigator to enter a patient quickly, errors can be made. For example, the investigator may tell the CRA that all laboratory values are within the required ranges and ask that the patient be entered. Based on the verbal information, the CRA completes the checklist and registers the patient. However, when the written laboratory results are received, the value for WBC is 3900 mm^3, and this makes the patient ineligible.

A checklist where actual values are requested can avoid this kind of problem. Example 2 is an expanded version of Example 1, allowing actual values to be recorded. It asks for confirmation of the eligibility criteria and collects specific values where relevant. It can also be used as a paper checklist which has to be filled out at the institution before registration.

EXAMPLE 2: An expanded version of the format in Example 1, allowing for collection of actual values.

Eligibility Checklist
Protocol: _____   Patient ID: _____   Institution: _____

Yes No

__ __ Is patient 15 years of age or older? Patient's Age in Years _____

__ __ Has the written pathology report been received?

__ __ Is there a confirmed histological diagnosis of colon cancer?

__ __ Is there evidence of metastatic disease?
         What is the date of surgery? __ / __ / __ (month, day, year)

__ __ Is the date of surgery more than 21 days and less than 60 days from today's date?
         When were the most recent hematological values obtained?
         __ / __ / __ (month, day, year)

__ __ Is this within 14 days of the date the patient will be registered?
         What is the patient's WBC? _____ (mm^3)

__ __ Is the WBC >4000mm^3?

Note: The answers to all questions in this section must be "Yes"

Yes No

__ __ Is the patient pregnant?

__ __ Has the patient had prior treatment for this cancer?

__ __ Has the patient any history of cardiac problems?

Note: The answers to all questions in this section must be "No"

Patient is eligible for the protocol if all above criteria are met

This format is preferable to that used in Example 1 if actual values are important in checking eligibility. It provides more information than Example 1 and ensures that important values are available at the time of registration.

Example 3 shows a format that would be appropriate for a computerized checklist. In this example, the computer is programmed to do the necessary calculations, reducing the number of questions that need to be asked. If programming resources are available to develop the necessary software, this option is the most efficient, for it reduces the number of questions that need to be asked. However, at the participating sites, the person completing the checklist must be aware of the allowable ranges for answers to the questions. These ranges could be printed on the checklist to make the completion easier.

EXAMPLE 3: Example of a format that is adaptable to computerized eligibility checks

---

Eligibility Checklist

Protocol: _____   **Patient ID:** _____   **Institution:** _____

What is patient's age in years? _____
(**Note:** Must be 15 years or older)
Is there a confirmed histological diagnosis of colon cancer? _____ (y/n)
Is there evidence of metastatic disease? _____ (y/n)
What was the date of surgery? __ / __ / __ (month, day, year)
(**Note:** Must be >21 days and <60 days from today)
On what date were the most recent hematologic values obtained?
          __ / __ / __ (month, day, year)
What is the patient's WBC? _____ (mm^3)
(**Note:** WBC must be >4000mm^3)
Is the patient pregnant? _____ (y/n)
Has the patient had prior treatment for this cancer? _____ (y/n)
Has the patient any history of cardiac problems? _____ (y/n)

This format reduces the number of questions being asked because calculations are done automatically. The computer calculates that the age, dates of surgery and hematologic tests, and WBC are within range for the protocol and checks whether the other answers are acceptable.

The examples given above are straightforward, where every question has to be answered for every patient. In some circumstances there may be a need to introduce conditional questions that are asked only depending on certain answers to other questions. For example, if a protocol allowed all patient under the age of 60 years to enter a trial and also allowed patients between the age of 60 and 70 to enter if they had a normal EKG, the checklist would have to be formatted so that the question about the EKG was asked only if the patient was between 60 and 70. This is most easily done by numbering the questions and introducing skips.

EXAMPLE: Checklist with Conditional Answers

1. Is the patient less than 60 years of age? _____ (y/n)

> If the answer is yes, skip to question 3
> If the answer is no, go to the next question

2. Is the patient between 60 and 70 years of age? _____ (y/n)

> If yes, has the patient had a normal EKG? _____ (y/n)
> If no, the patient is ineligible for this trial

3. Next eligibility question.

When checking eligibility for a patient, it is important that the protocol criteria be strictly enforced and that exceptions not be allowed. If the protocol is unclear or ambiguous, then a protocol amendment should be issued to rectify the problem rather than allowing eligibility exceptions. The revised criteria would then be applicable only to patients entered after it was introduced and should not be applied retrospectively to patients already entered. When a trial is audited, the protocol is the standard by which a case will be judged, and if the patient's characteristics do not comply with the protocol as written at the time the patient was entered, the case will be considered ineligible.

LOGGING RECEIPT OF DATA

When case report forms are received at the Coordinating Center, it is recommended that they be stamped with the date of receipt and logged as "received." This can be done in several ways. If there are only a small number of forms and cases, they can be manually stamped and entered into a logbook. If the volume is large, it is advisable to develop a more automated system or a system that is a combination of the two. The form/patient/trial identifiers can be keyed into a computerized database log, or if programmer support is available, barcode or optical scanning technology can be used. This involves ensuring that all case report forms are precoded with a barcode or identification label. To link the form to a specific patient, there also needs to be a barcode on the form that identifies the patient. This can be done by generating patient-specific forms, or by generating a set of barcoded patient identification labels when a patient is entered on study and sending them to the participating site to use on all forms for that patient. Logging receipt of forms simplifies the process of monitoring submission of forms in a timely way. If electronic forms are used with automated transfer of records to the Coordinating Center, the software should be programmed to maintain a database file recording the receipt and date of receipt of the records.

CHECKING FOR CORRECT IDENTIFIERS

Chapter 3 discusses the importance of including fields for identifiers on each form or page of a form. These identifiers can be for the protocol, the institution where the patient is entered, and for the specific patient. When case report forms are received at the Coordinating Center, it is important that they be checked for the presence of the appropriate identifiers before any processing is done. If forms have the wrong identifier on them and the information on them gets entered into the database for the wrong patient, these can obviously be a serious impact on the integrity of the trial.

The trial name or number should be on every case report form as well as the unique patient identifier that has been assigned, and for data that arrive by mail, the identifiers can be checked before the forms are separated from the envelope in which they arrived. However, in a busy office where a lot of mail is received, it can be difficult to maintain a system where the envelopes are always correctly matched with their contents, and it is therefore useful to request that the institution identifier be included on each form, either as a code or as a text string. If there are problems and the data cannot be matched with those already received for a patient, the forms can be returned to the participating institution for clarification. If the Coordinating Center wants to keep the return of forms to a minimum and wants to identify all that they possibly can, a central master list can be checked to see if the correct identifier can be assessed from the information available on the form. For example, if the patient's initials or hospital ID number are on the form, a cross-check can be done. For large trials it is common to introduce safeguards against use of incorrect identifiers. Check-digits can be built into the assigned patient identifier, and computers can then detect almost all errors in the recording or transcription of these identifiers.

CHECKING FOR DATA COMPLETENESS

It is important that all questions on the case report forms be answered, and it is the responsibility of the Coordinating Center to check that the submitted forms are complete. This check can be done either manually or using a computer. If the checks are done by computer, the data from the case report form must be entered into the computer before the check is done; if the check is done manually, it can be done either before or after data entry. In general, missing information needs to be requested from the institution that entered the patient on the trial. However, if the institution indicates on the form that the data are permanently unavailable (e.g., a test was never done at the relevant time point), then there is little reason to continue to request the information. It is recommended that a standard coding convention be developed to distinguish between data that are permanently unavailable and data currently "unknown" but requested from the institution. There may also be some missing data that can be handled centrally. For example, if the "day" field of a date is missing, it may be acceptable to routinely record the day as "15." Such rules need to be defined ahead of time and agreed on by the statistician and the Study Chair, and they may not be applicable to all date fields, only some. For example, it may be acceptable to use this convention for "date of birth" but not for "date treatment started."

RANGE AND FIELD TYPE CHECKS

Range errors are detected when the value in a specific field does not fall within predefined ranges of allowable values for that field. Likewise a field type can be predefined, and data can be checked to ensure that it is of the correct type. For example, an alphabetic character is not acceptable in a field that is defined as "integer." Range and field type checks can be done at the time of data entry and errors can be flagged. If the error is because the data entry operator mis-read the value on the form, the correct value can be re-entered at that time. However, if the value written on the form generates an error, then the error needs to be flagged and referred to an individual who can confirm or correct that value. If data entry is being done at the Coordinating Center, this type of error normally needs to be returned to the institution for clarification. If data entry is being done at the institution, then the error can be corrected on site before data are transferred to the Coordinating Center.

LOGICAL AND CONSISTENCY CHECKS

Logical and consistency checks should also be part of the Coordinating Center quality control program. When data come in, it is important not just to review the form that has just been received but to also check that form against all data previously received for that patient to make sure that data are consistent over time. For example, if a patient is started on protocol treatment on April 1, 1994, it would not make sense to have the most recent date known to be receiving treatment reported as January 10, 1994. In this instance, the year has probably been incorrectly coded as 1994 instead of 1995, but even such an obvious error as this should be confirmed with the institution in case the assumption over the cause of the error is incorrect. It is conceivable that the date is correct but that the wrong patient identifier was on the form.

MANUAL OR COMPUTERIZED CHECKS

Quality control checks can be done by visual review of the submitted forms, by computer programs run on the database generated for the trial, or as a combination of the two. Data recorded in "self-coding" forms (i.e., where boxes are filled in for each response or all responses are in a multiple-choice format) are most conducive to automated checks, while data that include text strings, diagrams, or narrative reports usually require visual review. Most trials are a combination of both types of data, and the Coordinating Center normally incorporates both kinds of checking into their quality control process. With the advent of optical scanning equipment, and the ability to store digital images of materials such as forms and X rays, it is probable that almost all checking will be able to be done electronically in future.

QUERIES TO THE INSTITUTIONS

Any part of the quality control process can initiate a query back to the participating institution, and a system needs to be developed to track queries sent and answers received. It is strongly recommended that all queries and responses be done by letter/fax/electronic mail rather than by telephone so that there is a permanent record of both the question and answer in the patient's record. In this way any data in the computer database can be supported by data on forms that are in the paper record at the Coordinating Center. If collecting information by telephone is unavoidable, then the person making the call and the person giving the information should both document the content of the call in writing and add the documentation to the patient's trial record.

At minimum, a query log needs to contain the patient/protocol identifiers, the date the query was sent, and the date that a response was received from the institution. The text of the query can be kept in the log, but if the actual query letter can easily be looked up, it need not be. Query letters can be written completely by hand, written with the help of computer software, or generated automatically after computer checks are run. The query itself, besides containing the same information as in the query log, must also have the name of the investigator and institution where it should be sent. Computer technology does allow for transmission of queries and response by fax or by electronic mail/file transfer. If all data items for a trial have a unique identifying number, it is possible for a computer-generated edit check or query list to be sent to the participating sites by the Coordinating Center with a list of the item number and the corresponding question. For example, a query list could have entries like the following:

	Edit Query	
Protocol: _____		**Patient ID:** _____
Form	Item number	Query
History	10	This item left blank—please provide answer
Pathology	22	Pathology Code invalid—please check

If a trial is being used for an application to a regulatory agency, it may be mandatory for the investigator to resubmit corrected case report forms if the query indicates that the original form was in error. In other instances, it may be adequate to respond to the query on the query form, sign the response and return. This will give the Coordinating Center sufficient data to modify the database. If revised forms are submitted, it is best to mark them clearly as *revised* and, if possible, to highlight the fields that have changed so that the Coordinating Center is aware of where the revision was made. Copies of all queries and responses should be kept as part of the trial record both at the institution and at the Coordinating Center.

REQUESTS FOR OVERDUE DATA

The Coordinating Center should develop a mechanism for reminding institutions when specific case report forms are due to be submitted. This reminder can be prospective and remind the participant *before* the form is actually due, or it can be retrospective and request the missing form after the due date has passed. To be useful, it has to allow for delays in mailing and processing at the Coordinating Center so that the participants are not constantly being asked for recently submitted data. The data request should clearly identify the protocol, the institution, the patient, and a list of all forms and responses to queries that are overdue. If reminders for data are prospective rather than retrospective, they can remind participants about special study requirements as well as case report forms that should be submitted. For example, if a blood sample has to be submitted for each patient after 12 months on trial, the data request could remind the institution prior to the due date to be sure that the requirement was met.

ASSESSMENT OF STUDY END POINTS

A critical part of central quality control is to evaluate the study end points and confirm that the institution has reported them correctly. The protocol should contain a section on the criteria to be used to measure the effect of the protocol treatment, and the institution should use these criteria when assessing the patient's outcome. A re-evaluation should be done at the Coordinating Center, checking to see that all appropriate tests were done and reported, and that the outcome was correctly assessed and accurately reported on the case report forms.

An example of an end point common to many trials is assessment of side effects due to treatment. Normally there will a grading scale used in reporting the type and severity of these side effects, and the Coordinating Center should review any supporting data submitted to check that the grades are accurately reported. Some side effects are graded based on laboratory test results, and as long as the actual results are available, these grades are easy to confirm and can be done either manually or by computer. Other side effects, such as "headache," are more subjective and more difficult to assess. A narrative description of the severity of the headaches and/or the medication taken to eliminate them is usually the best data available to confirm grading of this kind of toxicity. It is important that the grading scales to be used are included in the protocol document so that all participants follow the same criteria.

In some trials, materials may need to be submitted to a central location for review and assessment of endpoints. For example, photographs, bone scans, or Xrays may document outcome and be sent to a central reference or reading laboratory. Use of this mechanism does ensure a consistent and objective review of these materials, since the personnel who evaluate them usually know nothing about the individual patients and base their review solely on the technical assessment of the materials provided. Quality control procedures also need to be established for the Reference Centers to ensure that their procedures are consistent throughout the trial.

It is recommended that the database contain fields that record the results of the central assessment of key end points, either by Coordinating Center staff or, where relevant, by the central reference facilities. These summary fields are useful for a statistician in the analysis of the data because they ensure that the statistician uses the outcome data generated as a result of the objective quality control procedures that have been followed for all patients. If all quality control and assessment are done automatically, these variables can be calculated, and if the assessment and checks are done manually, an internal form could be developed to document the outcome. The following example shows a simple form that can be used for assessment:

EXAMPLE: Case Evaluation Form

Evaluation Form

Protocol Number: _____   **Patient ID:** _____

_____   Is patient eligible for the trial (y/n)
_____   Was treatment given according to protocol (y/n)
_____   Are sufficient data available to evaluate side effects? (y/n)
_____   Is patient evaluable for assessment of efficacy/response (y/n)
_____   Are there other problems with this case? (y/n)

It may be useful to expand this form so that the reason for problems can also be recorded. Such details will probably be study specific, and they will also make analysis and reporting much easier for that specific trial.

CLINICAL REVIEW

In many trials it is advisable to have data and outcome assessments reviewed by a specialist, either in addition to the Data Coordinator review at the Coordinating Center or in place of it. In some trials this review will take place at the end of treatment when all data have been received; in others there could be early review of the treatment compliance plan and/or ongoing review of data. For example, in trials that involve radiation therapy, there is often central review of the radiation treatment plan and treatment fields before the treatment starts to ensure that the radiation therapy will be given according to the protocol. This type of review requires rapid turnaround so that the reviewer can contact the treating radiation therapist before treatment begins if possible problems are identified during the review. To do this, there needs to be a mechanism for sending the necessary materials by express mail or courier, and there always needs to be a specialist available to do the reviews. Computer technology may simplify this process in future by allowing electronic transmission and review of images.

It is certainly advisable to have ongoing clinical review of all serious Adverse Events reported in a trial. The rate and type of events should be constantly monitored to detect unexpected patterns and types of events. Even events that the treating physician considers unrelated to treatment should be monitored centrally, since it is only by reviewing data from the entire study that these unexpected patterns can be seen. For example, if a physician enters a patient on a trial and that patient has a fall and breaks a bone, then the local physician will probably consider that unrelated to the protocol treatment. However, if there are several reports of broken bones across all patients entered, research may find that the protocol drug is having a previously undetected adverse effect and causing brittle bones, or that the drug makes patients dizzy and they fall. Phase I and II studies are done using small numbers of patients, so it is often in the early Phase III testing of new drugs that these unexpected side effects are detected.

In addition the Study Chair should at minimum review all "problem" cases on a study—cases where there are unresolved problems in assessing the compliance to protocol and the patient's outcome assessment, or cases where there were protocol violations. The Study Chair and the statistician need to decide how to handle these cases in the final analysis.

Having independent clinical review of all cases entered on a trial will add to the quality and integrity of the data and should be implemented when feasible. The evaluation form shown in the example above could easily be expanded to allow recording of the results of the initial review at the Coordinating Center and then final review and adjudication by the Study Chair. If the Study Chair is going to review any or all cases, a mechanism for the review process needs to be developed before the trial activates. The Study Chair could periodically visit the Coordinating Center to review all data on site, or if this is not practical, copies of all data received can be sent to the Study Chair either on an ongoing basis or with periodic mailing of batches of data. It is recommended that these reviews be done at minimum twice a year. Doing the review after the trial is completed will help to adjudicate on how cases will be treated in analysis, but it will not help to identify any ongoing clinical problems in the early stages of the trial.

FEEDBACK TO PARTICIPANTS

In any trial there will be errors in the completion of case report forms and problems with protocol compliance, so it is important that there be constant feedback to the participating institutions. If there is any indication that the institution does not understand the protocol or the instructions for the case report forms, the Coordinating Center should contact the institution and discuss the problems. It is also important to inform the institution about the results of the Coordinating Center's evaluation of a case so that the results can be challenged where disagreement occurs. For example, if the Coordinating Center codes a case as a protocol violation because one of the protocol drug doses was reduced for no apparent reason, the institution may be able to provide additional documentation that there was a valid explanation for the reduction and the assessment of the case can be reconsidered. The Coordinating Center should also notify the institution when it disagrees with assessment of data items on the forms. For example, the institution may report the date of treatment failure as the date that the failure was confirmed by diagnostic tests. However, the Coordinating Center may recode this date to reflect the date on which the first indication of failure was reported. This could have been patient symptoms or other subjective criteria. While this is a difference in coding convention, it does help the institution to submit consistent data on future cases. Such constant iteration between the institutions and the Coordinating Center will improve the quality of future data and can help to clear up misunderstandings.

EVALUATION OF PERFORMANCE

If a trial is in progress for several years, or if the same institutions participate in many trials conducted by the same Coordinating Center, it is useful for the Coordinating Center to evaluate the performance of the participating institutions at regular intervals. The performance review can include accrual rates, eligibility rates, protocol compliance rates, and timeliness and quality of submitted data. The evaluation could include an overall assessment of performance so that institutions get a clear idea of any areas where they need to improve. Assessments could include categories for commendation, approval, warning, and probation. The last two would include information about areas that need to improve before the next evaluation. Institutions that have persistent problems despite efforts to educate the staff may need to be dropped from participation.

QUALITY CONTROL AT REFERENCE CENTERS

As mentioned earlier in this chapter, it is important that there be quality control of the work done by any Reference Centers in a trial. Some quality control can be done by the Coordinating Center when it reviews the data sent by the Reference Centers, but there should also be quality control of the methodology being used within the center. The process will depend on what the Reference Center does but could include a reassessment of materials from random cases (selected by the Coordinating Center) to ensure that the results from the second review are the same as the first. It may also be possible to send some samples to another Reference Center for review to see whether there is consistency between reviewers. The Coordinating Center should be responsible for setting up this quality control process and for monitoring the results.

QUALITY CONTROL AT THE COORDINATING CENTER

The procedures in place at the Coordinating Center should also be subject to quality control. Options for ensuring accuracy of data entry are described in Chapter 5. Further quality control can be done by implementing a system where cases are reprocessed by a different Data Coordinator than the one who originally processed the data. The second review should be done "blind" with no access to the data generated by the first review. In other words, the second Data Coordinator reviews all the incoming data without reference to the database or to any notes made by the first Data Coordinator. The results of the second review can then be compared with the first. This process should not be seen as a punitive one but should be considered part of the routine quality control requirements for the trial. Quality control of clinical trial data is a complex process, and there are many conventions that have to be followed. It is easy for differences in interpretation to creep into the process, and having these periodic second reviews is another method of ensuring consistency in processing across all cases. It can of course also detect problems with the accuracy of Data Coordinators in following the documented procedures.

SUMMARY

The Coordinating Center quality control program is an essential component of a clinical trial. It should encompass all stages of the trial, from eligibility checking to evaluation of end-point data. A system needs to be in place for logging receipt of data and dealing with questions to the institutions. Clinical input into the data review and feedback to the participants are also important parts of the quality control process. In general, the sooner an error is detected, the more likely it is that a correction can be made. All types of checks should be done on an ongoing basis as data are received rather than batching forms and doing quality control at the end of the trial. If queries are sent to a site months or even years after a patient was entered and treated on a trial, the likelihood of retrieving missing data is diminished. Continuing education of participants will also improve the quality of data collected for a trial. Quality control procedures should also be implemented at the Reference Centers and at the Coordinating Center itself to ensure consistency over time.

CHAPTER 9

Data Management
and Good Clinical Practice

The importance of high-quality data management in the conduct of clinical trials has been increasingly emphasized with the unfortunate reports of cases of fraud in a small number of clinical trials. There should be adequate safeguards in place in any trial to ensure the integrity of the published data and that all trials are conducted to high standards of clinical practice. The standards that define the conduct of clinical trials are known as the Code of Good Clinical Practice (GCP). International agreement has been reached on the baseline standards, although implementation sometimes varies because of national regulations. The regulatory agencies that are responsible for the approval of new drugs before they become commercially available review the conduct of any trial used in an application for approval and ensure that Good Clinical Practice standards have been met. Data management procedures are very important in meeting the standards, and these standards should be met for any clinical trial whether or not it is being filed with a regulatory agency. This chapter will provide an overview of the requirements for Good Clinical Practice and give suggestions for implementation and monitoring both at the participating sites and at the Coordinating Center.

GUIDELINES OF GOOD CLINICAL PRACTICE

Good Clinical Practice represents standards by which clinical trials are designed, conducted, and reported ensuring that the patient's rights have been protected throughout the course of the trial and that there is confidence in the integrity of the collected data and the published results of the trial. The standards were developed as a result of tragedies in the past, such as the use of thalidomide in the early 1960s. The World Health Organization (WHO) initiated discussions to regulate clinical research involving humans, and the discussions led to an international agreement signed in 1964. This agreement, known as the Helsinki Declaration was adopted by the World Medical Assembly and addressed ethical issues in the conduct of clinical trials. The Declaration covers the rights of patients, and defines clearly the requirement for fully informed consent from the patient. The Declaration emphasizes the patient's right to withdraw at any time and that refusal to participate in the trial should not affect the patient's future treatment or have any negative effect on the patient's relationship with the physician. This standard must be met in any clinical trial.

Many countries have expanded on the requirements of the Helsinki Declaration and have regulations covering the development and testing of new drugs or treatment modalities. In the 1960s the United States developed procedures to be followed for testing of new drugs, and the FDA (Food and Drug Administration) introduced their Code of Good Clinical Practice in 1978. The European Union published revised guidelines for Good Clinical Practice among its members in 1987, and in 1995 many nations participated in the International Conference on Harmonization, held in Yokohama, Japan. This conference was held to discuss international standards and procedures for testing of new agents. While there are still variations between countries, the basic standards are agreed on. They protect the rights of the patient entered on a trial and set standards of practice to ensure the integrity of clinical trials. Whether or not a trial uses investigational treatments, or will be used as part of an application for approval of a treatment to a regulatory agency, meeting these standards of Good Clinical Practice help ensure that a clinical trial is conducted appropriately, meeting all ethical and regulatory requirements.

GOOD CLINICAL PRACTICE RESPONSIBILITIES

Both the Coordinating Center (representing the sponsor) and the local participating sites have responsibilities in ensuring that Good Clinical Practice standards are met. Both should be fully aware of their own responsibilities and ensure that systems are in place to meet them.

Good Clinical Practice Responsibilities at Participating Sites

At local sites where patients are entered on clinical trials, both the investigator and the Clinical Research Associate (CRA) need to be aware of the regulatory requirements for the trial prior to participation. The investigator is the person responsible for the clinical aspects of the trial and for the protection and clinical care of the patients entered on the trial. There may be multiple investigators at one location involved in a trial and normally one of these is identified as the Principal Investigator (PI). The PI assumes overall responsibility for the conduct of the trial at that site. The Clinical Research Associate is appointed by the investigator to assist in the administration of the trial at that site. In places where only one or two cases are entered on a trial, the PI sometimes assumes the responsibilities of the Clinical Research Associate.

As well as the care of the patient entered on the trial, the responsibilities of the investigator include cooperation with the sponsor of the trial to ensure that its obligations and responsibilities can be met. This may include providing access to medical records, answering questions about cases entered on the trial, and ensuring Good Clinical Practice guidelines are followed. The PI is ultimately responsibility for the accuracy and completeness of the data recorded on the case report forms, even if the forms are completed by the Clinical Research Associate. When trials are being used for an application to a regulatory agency for the permission to make the treatment commercially available for a particular indication, it is common for the sponsor (usually a pharmaceutical company) to require that the investigator sign a copy of all completed data forms and also a copy of the protocol document to indicate that it has been read and understood and that patients will be treated according to the protocol treatment plan. The regulatory requirements can be extensive and will vary depending on the trial being done and the country or countries where it is conducted, so it is important that participants fully understand their responsibilities prior to activation of the trial.

During the conduct of a clinical trial, the local site may have some or all of the following responsibilities:

1. *Ethics Committee Approval.* The local site must ensure that the protocol has been approved by a local Ethics Committee/Institution Review Board (IRB) *prior* to the entry of the first patient. Ethics Committee membership usually includes medical professionals and lay members, and the Committee usually enjoys a substantial degree of autonomy in interpretation and enforcement of the regulations.

   Submission of the protocol for approval by the committee may involve preparing a summary of the protocol and providing a copy of the patient consent form/information brochure. In the United States it is a requirement that the Ethics Committee receive a copy of the protocol and consent form approved by the sponsor. Any changes to key sections of the consent form must be justified and may need to be reviewed by the study sponsor. It may also be a requirement that the protocol be re-reviewed at regular intervals. In the United States re-approval of an ongoing protocol is required annually; in other countries the reapproval is requested every two years.

2. *Recruitment of Patients to the Trial.* Local sites are asked to participate in a trial primarily because of their ability to enter and treat patients on that trial. If a site does not enter the projected number of patients on to a trial, their right to participate may be withdrawn by the sponsor. Recruiting patients involves checking the eligibility of potential patients, explaining the trial to them, and obtaining their consent to participate.

3. *Obtaining Patient Consent to Participate.* Informed consent is the process of explaining the trial to patients and getting their agreement to enter the trial. It is an essential part of Good Clinical Practice to inform patients about the trial and the possible risks associated with the treatment. As part of the consent process, each patient should receive the following information:

- A description of the research plan, with rationale for the treatments being studied.
- Information about possible risks and benefits.
- A summary of any procedures or requirements that go beyond standard practice.
- Information about the alternative treatment options.
- Information about possible costs or compensation.
- Name and telephone number of both the responsible physician and an individual to contact if there are problems or questions.
- Assurance that confidentiality will be maintained, but that certain people such as the Coordinating Center/sponsor will have access to the data collected for the trial and may also have access to the medical records, depending on the laws in that country.
- Information that the patient has the right to refuse participation without affecting future clinical care or their relationship with their clinician.
- Assurance that the patient has the right to withdraw from the trial at any time, again without negative effect on their future medical care.

  The written information given to patients for signature can vary in format. In the United States the patient must sign a consent form that contains all of the above elements. It can often be several pages long and be quite intimidating to a patient. In other countries the patient is given an information brochure with the above information, but the consent form actually signed is short, usually only one page. It is usually required that a witness sign the consent form in addition to the patient, and there may also be a requirement for the physician to sign. Most Ethics Committees require that the consent form be prepared in a specific format, and these formats can vary substantially from one place to another. Some countries allow verbal consent to obtained as long as it is documented in the medical record.

4. *Collection and Recording of the Data Required by Protocol.* The data required for completion of the case report forms for the trial must be collected and transcribed on to the case report forms. Source documentation for these data also needs to be maintained so that the data on the case report forms can be verified and substantiated.

5. *Reporting of Adverse Events.* Trial sponsors require that all serious side effects of protocol treatment be quickly reported. The protocol should contain instructions on the type of event that must be reported, and the method and schedule for reporting. It is the local site's responsibility to report all such events. It is usually a requirement that these events also be reported to the local Ethics Committee/IRB.

6. *Ensuring Protocol Compliance.* It is the responsibility of the Principal Investigator to ensure that all patients are treated according to protocol. This includes following the protocol treatment plan, ordering all required tests on schedule, and making the required assessments when appropriate. If a site has persistent problems with compliance, its participation may be terminated.

7. *Ordering, Storage, and Administration of the Study Drug.* When a clinical trial involves administration of medication, the local site must ensure that the drug is handled appropriately. The protocol should contain instructions on how to obtain the study drug once a patient has been entered. It is the responsibility of the site to order the drug and maintain records that document the receipt and dispensing of study drug. For double-blind studies, when necessary, there should be documentation that the pharmacist followed the required procedures for ensuring that the treatment remained blinded to the investigator and patient. There also need to be records that document the destruction or return of unused drug. There may be special regulations that govern the storage and handling of trial drugs, particularly if they are investigational (not yet approved for this indication). These requirements could include maintaining separate supplies of drug for different protocols even if the same drug is used for more than one protocol. If two protocols are

active and using the same drug, it is important to check whether drug from one protocol can be "borrowed" and "paid back" from one protocol to another or whether the supplies must be kept completely separate. If drug is being supplied by a sponsor, it is important that the drug supplied be used, even if the drug is available commercially. It is also important that if a commercially available drug is provided, that it not be used for nonprotocol patients. The pharmacist, or individual responsible for the storage and dispensing of the drug, should be aware of all regulations for that drug and protocol.

Good Clinical Practice Responsibilities of the Coordinating Center/Sponsor

The sponsor of the trial is responsible for monitoring the progress of the trial and ensuring compliance with regulations. Sponsors can supervise or conduct this monitoring themselves or assign the responsibility to a Coordinating Center. This latter model is the one primarily used in this text. The responsibilities include all or some of the following:

1. *Confirming Ethics Committee Approval at each Site.* The Coordinating Center is responsible for ensuring that all regulations are met, and this includes making sure that participants have the Ethics Committee Approval of a trial prior to entering their first patient. This can be done by collecting copies of the approval forms/letters or by asking whether approval has been obtained at the time a participating site enters its first patient. If periodic reapproval is required, the sponsor will need to develop a system for monitoring this. It may also be necessary for the Coordinating Center to ensure that the composition of the Ethics Committee's membership is appropriate before the trial starts.

2. *Confirming Informed Consent.* The Coordinating Center must also have a system for ensuring that before a patient is registered/randomized, informed consent has been obtained from the patient. This can be done by collecting copies of signed consent forms after registration, or by asking at the time of registration of a patient whether a consent form has been signed.

3. *Source Verification of Submitted Data.* It is the Coordinating Center's responsibility to confirm the accuracy of data reported on the case report forms. This is normally done by regular visits to the participating sites to review source documents. In some trials that are being used for filing with a regulatory agency, there is usually verification of

source data for all cases. In other situations the percent of cases reviewed will vary.

4. *Screening Personnel.* The Coordinating Center must confirm that the participants in a trial are appropriately qualified to fulfill the requirements of the protocol and are not currently barred from participation in trials by any oversight body. This normally involves the collection of current Curriculum Vitae from participating researchers.

5. *Quality Control and Computerization of Case Report Forms.* The Coordinating Center is responsible for quality control of submitted data. Procedures are described in more detail in Chapter 8 and include review of all submitted data for completeness and consistency and generation of queries to participants to ask questions or request missing data/overdue forms.

6. *Ongoing Monitoring of the Safety of the Treatment(s).* It is important for the trial sponsor to be sure that there are not unexpectedly high rates of Adverse Events (AE) in a trial, and the Coordinating Center must be sure that there are mechanisms for the rapid reporting and review of all Adverse Events. This should include review of all reported Adverse Events and also a review of submitted case report forms to ensure that all such events are being appropriately reported in a timely way and that the incidence of side effects does not exceed the expected rate.

7. *Analysis of the Trial and Reporting of Results.* A well-designed trial will specify the analysis plans in the statistical section of the protocol, and analysis should be done according to that schedule. The Coordinating Center should develop the time frames for analysis and ensure that there is a mechanism for review of the results. Phase III trials are usually kept blinded to outcome until the data are considered mature and definitive. This means that until sufficient data have been received to ensure that the results are statistically meaningful, the results will not be made known to participants. Phase III trials are usually monitored by a Data Safety and Monitoring Committee, and that committee will make the decision when trial results can be made public.

IMPLEMENTATION OF GOOD CLINICAL PRACTICE AT PARTICIPATING SITES

As described, the participating site has several responsibilities in complying with Good Clinical Practice requirements. This section provides information on those responsibilities and provides some ideas for ensuring that requirements are met. Additional suggestions for local data management procedures can be found in Chapter 7.

Source Documentation

Complete and accurate source documentation is critical for a clinical trial, particularly one that is being sponsored by a pharmaceutical company and that may be used in an application to a regulatory agency such as the FDA in the United States. Good Clinical Practice requires that data submitted on case report forms can be verified by review of source materials at the local participating site. There are different kinds of source materials. The main one is the patient's medical or clinical record. This is an official hospital/clinic chart that contains all information about the patient's medical history at that hospital/clinic. The content and format of the medical record are often mandated by a Medical Records department, and it may be difficult to keep all necessary trial records in this document. If this is true, a trial chart should also be maintained for any supplementary data that are not part of the official medical record. Copies of all case report forms submitted to a sponsor or Coordinating Center should be kept, as should copies of any data queries sent to the site by the Coordinating Center.

Organization of records into sections makes it easier to locate information when needed. For example, sections could be made for physician notes, nurses notes, prescriptions, operative/pathology reports, laboratory results, and results of X rays/scans. Within a section, records should be kept in chronological order, usually with the most recent record first.

There are usually regulations that govern the extent to which protocol treatment can be given totally or in part at a location other than the site that entered the patient on the trial. Usually investigational agents may only be given at the site where the trial team is located, but occasionally it is acceptable for some protocol treatment or monitoring to be done at a different location. For example, interim laboratory tests may be done by the patient's local physician rather than at the trial site. Before allowing any treatment or monitoring off site, check the regulations that apply to the specific trial.

If some parts of a protocol treatment are given at another hospital/clinic, it is important that copies of the relevant information be collected and maintained at the site that entered the patient on the trial. For example, if surgery is done at another hospital, copies of the operative and pathology notes should be requested. It is strongly recommended that any outside data be requested on an ongoing basis to ensure that the trial record is complete. If at all possible, information about the fact that the patient is on a clinical trial and about the data required for the trial should be given prospectively to the clinician who is treating the patient off site. This will make it more likely that the required data will actually be collected and recorded.

Other types of source materials include films such as X rays, CT scans, and EKGs. Pathology slides may also be considered source materials for specific trials, and these should be available for review by the study sponsor when required. It is particularly important to retain materials that were obtained at baseline (when the patient was entered on to the study) and materials that document response (or failure) on protocol treatment. Some trials will require that these materials be submitted for central review, and they are then retained centrally. In other studies it will be the responsibility of the participating sites to ensure that they are available for review when requested.

It is important to get information from the sponsor about the length of time that trial-related records must be kept after the end of the trial. Regulations about archiving trial data vary from country to country and could also vary, depending on the type of trial. In the United States original records need to be retained for a minimum of 7 years from approval of the treatment by the FDA. The European Good Clinical Practice guidelines recommend that trial data be kept for a minimum of 15 years. It is therefore important to be sure that the trial data can be easily identified and will be retained regardless of local medical records practices. It would be beneficial to discuss trial requirements with the Medical Records Department to see whether a system can be developed to easily identify trial patients and record the earliest date on which records can be destroyed. If the Medical Records Department are not willing to retain the record, then the Principal Investigator is responsible for making sure that a complete copy of the data relevant to the trial is kept in the trial record.

The data can be stored electronically, provided a backup (electronic) copy also exists and a hard copy can be generated if necessary. This means that microfiche or electronic image storage is acceptable as long as a paper copy can be printed when needed. If paper forms are being archived, it is important to be aware that faxes and NCR copies fade over time and that it is best to make a copy of these before storing.

Recording Required Data

While it is important that protocol requirements be followed and that all necessary trial data be collected, in everyday clinical practice it is hard for the physician to remember everything that needs to be done. Compliance can be improved if the Clinical Research Associate develops tools to assist the clinician in the management of the trial patient. Copying either the entire protocol or relevant sections of the protocol (treatment administration section, protocol test schedules) and placing them in the medical record for the physician is useful. However, these sections can be lengthy, and the physician may not always have time to review them carefully. There are also potential problems if the protocol changes and the copies have to be updated.

If the Clinical Research Associate can prospectively identify the patients who will be seen on a particular day, they can help the clinician by reviewing the protocol and the patient's record and making a list of tests that need to be done and data that must be recorded at the visit. For example, if side-effect data have to be collected, the Clinical Research Associate can remind the physician to ask the patient what side effects they have experienced since the last visit. Specific tools could be developed for data that have to be collected and recorded at each visit. For example, if tumor measurements are being monitored for a cancer patient, it is important that all sites be followed at evaluation time points. If some measurements are obtained by physical exam the only source documentation that is possible is a notation of measurements written by the physician. To ensure that the measurements are done and recorded, the Clinical Research Associate could develop a tumor measurement form which lists the sites to be followed and has space to record the dates and the measurements.

EXAMPLE: Worksheet for Recording Required Data over Time

SITE	DATE:	DATE:	DATE:	DATE:
	Measurement	Measurement	Measurement	Measurement
Supraclavicular Node				
R. Axillary Node				
Liver				

This simple form allows collection of data over time and ensures that the correct sites are followed. Because it has multiple observations, it also allows the physician to assess response or progression of disease. This form could be expanded to record other sites of involvement that are monitored by radiological techniques rather than physical exam. If so, another column should be added to record "Method of Evaluation." Similar tools can be developed for any clinical trial. If such forms are developed for recording source data, the physician who completes the form should sign/initial the form. More details of data management tools are given in Chapter 7.

It is also important to ensure that any grading scale or definition that is in the protocol is used appropriately in the assessment of the patients progress. Keeping a copy of these definitions in the patient's record makes it easier for them to be consistently used.

While these kind of systems do require time and effort on behalf of the research team, it is time well spent, for it ensures that the data on the case report forms will be complete and that source documentation will be available. Monitoring visits by sponsors are therefore greatly simplified, and the number of queries from the Coordinating Center will be greatly reduced.

Completing Case Report Forms

When completing case report forms for a clinical trial, there are procedures that should be followed to comply with Good Clinical Practice.

1. *Read and Follow All Instructions Carefully.* Instructions may be written on the forms themselves (sometimes on the reverse side), or there may be a Trial Manual with documentation and instructions. If there is a question about how to complete a form, it is advisable to call the Coordinating Center to clarify.

2. *Write Legibly on the Form.* Always use permanent ink and not pencil or erasable ink. Some trials require use of a certain color of ink in completing the form so that copies can be made easily.

3. *Submit Original Forms When Requested.* The original forms (not copies) should usually be submitted to the Coordinating Center, with a copy kept at the local site. It is easier for reviewers to read originals than copies, and most sponsors require receipt of the original forms.

4. *Write Answers Inside the Space Provided.* Do not write outside or over boxes provided on the form. If additional information must be provided, then use the "Comments" section of the form (if there is one), write clearly in any blank space on the form, or write the comments on a separate sheet of paper. If using a separate sheet, make sure that the correct patient and trial identifiers are written on the sheet in case it gets separated from the form. Do not write comments that cover up responses already written on the form.

5. *Answer All Questions.* If an answer is unknown or not available, that information should be provided to avoid a query from the Coordinating Center. If there is no code on a form for "Unknown" or "Unavailable," leave the space for recording the answer blank, and write

"unknown" or "unavailable" to one side. Do not write the words in the space provided for the coded answer.

6. *Use the Units of Measurement That are Requested on the Form.* If results are available in a different unit and the conversion algorithm is not available, do not write the value with the different units in the response field. Instead, write the value and units in a space on the form, indicating which question they refer to. Conversion can then be done at the Coordinating Center, and the response field completed. Be aware of any decimal points that are preprinted on the form, and ensure that the value is correctly recorded.

7. *Use Appropriate Mechanisms for Corrections on Forms.* If an error is made when completing a case report form, *never* use correction fluids of any type to erase the error. Instead, the erroneous information should be crossed out with a single line through it, and the correct response written above/to the side. The correction should be initialed and dated to indicate who made the correction and when.

8. *Check Forms Before Submission to the Coordinating Center.* Each case report form should be checked for completeness and correctness before submission to the Coordinating Center. If required, the Principal Investigator and/or the Clinical Research Associate should sign the completed form.

9. *Ensure that ID Information is Correct.* The correct patient/trial identifiers must be recorded on every page.

10. *Follow the Correct Schedule for Submission of Forms.* Complete and submit case report forms according to the schedule in the protocol.

Preparing for a Monitoring Visit/Audit

Trials that are sponsored by the pharmaceutical industry, a clinical trials cooperative group, or a government agency may require some degree of source verification. In other words, for a certain percentage of cases entered on the trial, there will be monitoring by the sponsor or a representative of the sponsor to ensure that the data submitted oncase report forms can be confirmed by the data in the patient's medical record at the local institution that entered the patient. As well as being audited by the sponsor, if the data from the trial are used for an application to a regulatory agency, an institution could also be audited by the regulatory agency.

The recent visibility of cases of fraud in clinical trials has led to an increase in the number of audits that are done, and participants in a trial should be prepared for them. Following the steps described for maintaining source documents and ensuring that all trial data are collected will help to make an audit go smoothly. Most sponsors will notify the institution ahead of time about the date of the audit and the cases selected for audit. However, agencies such as the FDA in the United States can request an audit without prior notification.

Maintaining a "Trial Master File" is sometimes required, but even if it is not, it is a good idea to ensure that the necessary trial documents can be easily located. A Master File would normally contain the following:

- Copies of the protocol and all subsequent amendments/revisions (if the trial requires that a copy of the protocol be signed by the investigator, the signed copy should be maintained in this file).
- Copies of blank case report forms. If there have been changes to the forms during the trial, the original and revised versions should be maintained for reference, along with dates of implementation.
- Blank consent forms and a copy of all signed consents for patients entered on the trial.
- Blank forms for reporting Adverse Events and copies of all actual Adverse Event reports that have been made for patients entered by this local site.
- Log of all patients entered on the trial along with the assigned patient identifiers.
- Information about ordering, preparing, and storing drugs.
- Documentation of normal ranges for the laboratory tests and certification/accreditation of the laboratory being used.
- Copies of Ethics Committee approvals and re-approvals of the protocol.
- Copies of correspondence about the protocol.

Not all of these documents will be required for every trial but maintaining those that are required in a central file makes it easier to locate the information for ongoing trial management and for audits.

There is a lot of preparation that can be done before an audit to make the process go smoothly. The following guidelines outline the steps that an institution can take to prepare for an audit:

1. Once the list of cases is known, the institution must collect all source documentation for those cases. The Coordinating Center or sponsor will normally tell you what materials will be required. Sometimes reports of radiographic materials will be sufficient rather than having to collect the actual materials. If the sponsor has not provided information

about what should be available, it is advisable to ask for that information. Source documentation will include the official medical record plus any supplementary records that are kept for trial patients. Original laboratory results will normally be required.

2. Once the records have been assembled, the Clinical Research Associate at the institution should go through the records to make sure that nothing is missing. This can be done by checking the data on the case report forms that have been completed with the medical record.

3. An audit can be done more quickly if the relevant sections of the medical record are "flagged" ahead of time. The Clinical Research Associate can insert markers for pages of the record that contain information relevant to the protocol. For example, mark the section where patient history can be found or where treatment data can be found. When the auditors arrive, the system for flagging can be explained to them, thus making it easier for them to locate the data that they need. If this marking can be done prospectively, it becomes much easier to prepare for an audit. This would mean that every record for trial patients would have ongoing markers placed in the record. The markers could be identifying tabs or colored labels stuck on relevant pages. This would make any trial record easy to follow at any time.

4. The Clinical Research Associate should also make sure that all required case report forms are complete and current prior to the audit.

5. Regulatory documents should be assembled. This includes the Ethics Committee approval (and re-approvals) of the protocol(s) being audited and copies of the signed consent forms for the patients being reviewed. The auditors may also want copies of normal ranges for laboratory tests and copies of certification/accreditation for the laboratories being used. If a Trial Master File is maintained, it should contain all necessary supporting information for the trial.

6. All key personnel involved with the trial at that site should be notified of the audit and asked to be available to answer questions. At minimum, the Principal Investigator must be available. At the beginning of the visit, either the PI or the Clinical Research Associate should explain the organization of the records to the auditors so that they can more easily locate the necessary information. The audit team will normally want to discuss the findings of the audit with the PI at the conclusion of the audit. This "Exit Interview" is important because it allows the PI the opportunity to clarify any misconceptions that may have arisen, and it may also provide additional information not initially made available. Audits are not meant to be punitive (nor should they be), but rather they

are fact-finding visits to improve the accuracy of the trial data. As such, any clarifications can only be beneficial and should always be encouraged.

7. If the pharmacy procedures are being reviewed, the pharmacist must have the trial drug logs available. These logs should document receipt, dispensing, and current inventory of study drugs. If drugs have passed their expiration date, the protocol usually indicates whether they should be destroyed or returned, and compliance with this will also will be checked. The auditors will review pharmacy procedures to ensure that the drug storage and handling facilities are appropriate and that there is adequate security to restrict access to authorized individuals.

8. A quiet room with adequate table space should be available for the auditors to work in. If any study materials other than records are being reviewed, the appropriate equipment would also need to be available. For example, if X rays/scans are to be reviewed at the audit, there should be a view box in the room.

These guidelines apply to audits where case report forms have been submitted to a Coordinating Center or trial sponsor. Most of them also apply to trials where the sponsor sends monitors out to the sites to go through the source documents and to complete the case report forms on site. Table 9.1 shows a basic checklist that could be used when preparing for an audit. The contents of the checklist can be varied depending on the trial being audited. This kind of checklist could be prepared by the Coordinating Center and circulated to a participating site prior to a scheduled audit.

GOOD CLINICAL PRACTICE IMPLEMENTATION AT COORDINATING CENTER

The Coordinating Center for a trial (or the sponsor) is responsible for implementation of Good Clinical Practice requirements at all participating sites and for monitoring compliance throughout the trial. This section includes information about these responsibilities and suggests ways to fulfill them.

Quality Control/Quality Assurance

The Coordinating Center is responsible for quality control of data submitted on case report forms and this activity is covered in detail in Chapter 8. The process will involve review (manual, electronic, or a combination) of the submitted data. Checks will be made for completeness and consistency, and

Table 9.1. Sample Audit Checklist.

AUDIT PREPARATION CHECKLIST	
Task	**Check When Done**
Make sure that all required records will be available from Medical Records	
Notify Coordinating Center of problems with availability of any records	
Organize copies of all submitted forms— make sure that forms submission requirements are up to date	
Go through records and forms and place markers in relevant sections of record	
Identify any problems with case	
Have all regulatory documents available for review	
Organize meeting room for auditors—make sure there is sufficient space to work	
If X rays are to be reviewed, arrange view box	
Make sure that Principal Investigator and other relevant personnel will be available	
Organize visit to pharmacy (if relevant)	

queries will be sent to the participating site if problems are detected. The Coordinating Center is also responsible for requesting overdue case report forms. This routine quality control should be done for all cases entered on the trial. The Coordinating Center should have consistent procedures for handling corrections to submitted data. These could involve asking the sites to submit revised case report forms with corrections. Alternatively, the Coordinating Center might accept answers to queries to be written on a separate form and submitted as documentation of the correction. Whatever mechanism is used, it is recommended that data be collected on paper (or electronic screens) and not by phone. All data items entered into the Coordinating Center database should be verifiable on review of the paper records received.

Organization of Audits

The Coordinating Center will be responsible for organizing and conducting the audits of participating sites. A schedule for ongoing visits must be prepared, and a system developed for notifying sites of an impending audit within an appropriate time frame. There needs to be a mechanism for selecting the cases to be audited. Audits could be done of all cases entered, or only a subset, depending on the study requirements. If a subset is being reviewed, the selection of cases can be entirely random, or a group of cases could be defined such as all cases with serious Adverse Events or all cases where the patient has died. To ensure a consistent standard of auditing at all sites, training programs and instructions for auditors need to be developed. Decisions need to be made about what data will be checked. It could be all data or a subset of key data such as eligibility criteria, baseline values, treatment administrations, side effects, and outcome data. Regulatory documents should always be part of the audit review process and, if appropriate, review of pharmacy procedures.

After an audit has been done, it is the Coordinating Center's responsibility to prepare a report and circulate it to appropriate people, including the Principal Investigator at the audited site. The PI should respond to the audit findings and make changes in local procedures if problems are found. The finalized reports are usually forwarded to the sponsor.

Monitoring of Compliance with Regulations

The Coordinating Center is responsible for making sure that all participating hospitals comply with required regulations. This is done in part through the audit process, but because audits are usually done only after patients have been entered and treated on a trial, it is advisable for the Coordinating Center to develop a mechanism for ensuring compliance before a patient is entered on trial by the institution. This could include collecting copies of Ethics Committee approvals and signed consents, or at minimum requiring the institution to notify the Coordinating Center that both have been obtained prior to registering a patient.

Training and Education

The Coordinating Center is responsible for training and educating participants in a trial. This can be done by preparing and circulating manuals documenting procedures and requirements for the trial. Pharmaceutical Companies often require that an initiation visit be made to each participating site before they enter patients on the trial. The purpose of this visit is to ensure that the staff at the institution understand the requirements of the trial, that any special equipment is in place and working properly, and that all regulatory requirements are understood and met. While this is usually done at a visit to the actual site, it can be done by holding a meeting of all potential investigators and their support staff. More details about training can be found in Chapter 12.

In general, the Coordinating Center should be prepared to answer questions on an ongoing basis and provide additional training if needed.

WHAT TO DO IF FRAUD IS SUSPECTED

The falsification of data for a clinical trial is a serious offence and one that can affect many future patients who may receive the treatment under study in a trial. There are always errors in data collection because the people who collect and process data are human beings and can make mistakes. The quality control procedures described throughout this text can help to overcome these human errors, and any erroneous values that are not detected after stringent quality control are usually unlikely to affect the overall results. However, deliberate misrepresentation of data is harder to detect during routine quality control, and it can usually only be detected during an on-site audit and review of original medical records. Even then, if someone is systematically providing wrong values, the records may not contain evidence of the original data.

If you are ever placed in the unfortunate position of suspecting that data are being falsified, it is essential that you report it to an appropriate person so that the situation can be thoroughly investigated. Normally the report would be made to a senior member of the Coordinating Center staff or to the trial sponsor. Be sure of your facts before taking such a step and be prepared to provide the information that led you to your suspicion of fraud. Taking such a step can leave you in a vulnerable position especially if the person you suspect is your supervisor. Discuss implications of your position with the person to whom you make the report. They may be able to provide some level of protection while the issue is being investigated.

The Coordinating Center should have clearly defined steps to be followed should a report of fraud be made. Such a report should be brought to the attention of the Head of the Coordinating Center, and that person would then be responsible for further actions. This would normally include immediate notification of the sponsor of the trial, review of data on any cases entered by the person in question, discussions with that individual and with others at the hospital/clinic where the person works. It is a situation that can have many repercussions but one that must be addressed in an expeditious manner so that the integrity of the trial can be assured.

SUMMARY

Good Clinical Practice in the conduct of clinical trials ensures that the interests of the individual patients are being protected and that results of a trial can be validated. Compliance with the regulations does increase the workload at the participating site, as well as the responsibility of the Coordinating Center. Tools can be developed to ensure that trial requirements are met, and it is important to monitor compliance with the regulations on an ongoing basis. Not all of the Good Clinical Practice guidelines will be appropriate for every clinical trial but most are appropriate practice for the conduct of any clinical trial.

CHAPTER 10

Software Tools for Trials Management

In every clinical trials application where a computer is being used, the users of the computer system will want to develop software to help them manage the trial efficiently and effectively. There are many tools that can be developed, and this chapter describes some that can benefit a clinical trials group. The benefits of the application, however, must be weighed against the resources needed to develop, use, and maintain the software.

DATABASE UPDATE

If data are captured on data collection forms and processed centrally, there must be a mechanism for entering that data into the computer and updating the data values into the trial database. Chapter 5 describes possible methods of data entry including remote data entry at sites with transfer of flat files to the central office, central data entry into flat files, or direct entry into the database itself. If either of the first two options is used, software needs to be developed to take the data from the flat files and put it in the correct location in the clinical trials database. There are several processing steps that can be included in an update program, and decisions need to be made about the complexity of the design of the program for a particular trial.

At minimum, the update process should check to ensure that data values are within the correct ranges for the field and that data fields are of the correct data type (e.g., a date field contains a date, or an integer field does not contain alphabetic characters). These checks can also be done at time of data entry, but a professional data entry operator is usually not qualified to interpret the error messages and make corrections to the data. It is often more effective to do these checks at the time of database update, generating a list of errors that need to be checked by a data coordinator.

Decisions need to be made about what to do with records that generate update errors. If there are multiple errors in one record, the record should

157

probably be rejected and not updated into the database. Likewise, if there are errors in key fields, the record should be rejected, since it could lead to retrieval errors later. However, if there are only one or two errors and they are not in key fields, either the record can be updated in its entirety with warning messages printed out about the errors or the correct parts of the record can be updated with a "pending review" value filled in the fields that generated errors. The error types and levels need to be determined prior to entering any data into the database, and trial-specific error checking programs need to be written and thoroughly tested. All data entered into the database should go through all defined checks. They should never be bypassed.

The output listings of error messages must be clear and easy to read by a data coordinator. They should identify the record in which the error was detected, the field(s) in error, the value that is in the field, the type of error, and the action taken. The following is an example of the type of error message that may be generated by an update program:

> **Update Error Message**
> **Record Number: 10**
> **patient number: 12345**
> **field name/ value: Date of Birth entered as 15/15/96**
> **error: Value of 15 for Month is invalid**
> **action: record accepted, field updated with null value**
>
> **Record Number: 25**
> **patient number: 12347**
> **field name/value: visit number entered as 15**
> **error: upper allowable range for field is 10**
> **action: record rejected**

These range/type checks should be done on all data entered into the database, whether at data entry or at time of update into the database. A dictionary-based DBMS usually allows the definition of field types, lengths, and allowable ranges. The DBMS may automatically do these checks for you, or you may have to write special purpose routines to ensure that the error listings are meaningful and accurate. As well as generating a list of error messages, an update program should generate a list of successful transactions so that all update transactions can be confirmed.

CALCULATED VARIABLES

The data update program may also be programmed to use the data in the input records to create calculated variables for inclusion in the database. For example, "time to failure" may be a data item that will be used frequently in reports of a trial. However, this value is usually not collected as a specific data item. It is usually calculated by using the "date of entry" on to the trial (or another relevant date) and the "date of failure." To simplify the generation of reports on the study, the update program could be programmed to automatically calculate the time to failure and to enter the resultant value in the database. The logic for the calculation would need to be defined and would normally be as follows:

> If "date entered" is present in the database and "date of failure" is in the transaction record being updated, then check to see if dates are valid and that "date entered" is earlier than "date of failure." If both dates pass these checks, then calculate the difference between the two dates and store the resultant value as number of days between the two dates.

There would also need to be some provision for corrections being made to either date after initial values for both had been entered in the database. Any change in dates would require the above check to be reapplied to the new date value(s). In this instance a warning message should be printed whenever the calculated variable is changed so that there is a second check that the update was an appropriate one.

There are other examples where a value would be calculated at time of each follow-up and would be expected to change every time a new follow-up visit was reported. An example is when the number of days on study change with every update of a new contact date. In this instance a change in the calculated variable would not require generation of a warning message unless the new value was *less* than the previous value.

These variables would be calculated every time a relevant new data value is included in an update transaction record. The calculation in the examples given above would normally be done when either the "failure date" or the "last contact date" was included in an update record. The "date entered" on study would usually have been entered in an earlier update transaction. The calculated variable would therefore require access both to the data in the update transaction record and to a value already stored in the database. For this reason, it may be more efficient to calculate variables with a program that runs after the update transactions have been applied, rather than retrieving existing database field values during the update run. This will depend on the capabilities of the software package being used and on the programming skills required and available.

The examples given here show the kinds of checking/calculating that may be done by an update program. The actual checks/calculations for a trial will obviously depend on the trial and the data being collected.

LOGICAL/EDIT CHECKS

The range and field type error checks described above are standard and really represent the minimum checking that should be done on data prior to update into the database. There are many more extensive checks that can be done either during or after an update. These logical (or edit) checks will have to be defined and programmed for a specific trial, since the relevant rules are usually different from one trial to another. Like the calculated variables logical checks are usually defined by interaction between two or more variables in the data set. These types of checks are usually most efficiently done outside the update program using a stand-alone program that runs against the entire database.

While logical checks will vary from trial to trial, there are some that will be the same across many trials, and some that will be very similar with only data values changing. Each trial could have a separate program written defining the checks for that trial database. However, in an environment where many trials are done over time, this is not efficient use of programmer time. It is suggested that in these environments, a data-driven checking system be developed. In a data-driven system, the checks for each study would be defined in a data-file formatted according to pre-set definitions. This file would then be accessed and processed by one program capable of reading and processing any data file in the required format. While the initial software development will be more extensive than a single program defining checks for one trial, the long-term savings can be substantial. This approach also means that new studies can be activated without programmer intervention as long as the data files can be accurately specified. To illustrate the difference, consider the following examples of two checks—one to ensure that the registration date is earlier than the date treatment started, and one to ensure that the age limit for a trial is met. In trial one, the lower age limit is 30; in trial two, the upper age limit is 75. The programming language used in the example is not a formal language but is used to show the type of statements that would be needed:

EXAMPLE 1: Programmed Logical Check

Date Check

if date_rx_started LT date_entered, print error

Age Check Trial 1

if age LT 30 print error

Age Check Trial 2

if age GT 75 print error

A program would have to be written for each of the two trials, with program statements equivalent to those shown here.

EXAMPLE 2: Data-Driven Logical Check

PROTOCOL 1 DATA FILE

Field Name	Test	Outcome
date_rx_started	< date_entered	error
age	<30	error

PROTOCOL 2 DATA FILE

Field Name	Test	Outcome
date_rx_started	< date_entered	error
age	>75	error

In Example 2, one program could be written to interpret both data tables and to implement the logical check system. The program code would be the same for both trials, and only the content of the table would be different.

These examples are short and very simple, but the difference in approach for environments where many large trials are done could lead to significant differences in staffing. If all logical checks were programmed using separate programs for each trial, a larger programming staff will be needed to implement the check programs. Also the maintenance burden is higher, since there are many more programs to maintain than with the second system. The need for programmer support before activating a trial could lead to delays in implementing new studies. With example two, the data tables describing the checks can be written by a knowledgeable Data Coordinator, so they do not need a programmer to write new code. This means that new trials can be implemented more quickly, and the number of programs being maintained is greatly reduced.

The extent of logical checking done electronically will depend on the overall quality control system, the method of data collection and the resources available for defining these electronic checks. It is efficient to use the computer to do as many checks as possible. Types of checks that can be implemented include checks for missing data, checks for possible wrong patient identifiers, checks for dates out of sequence, checks that protocol treatment was correctly administered, and checks for patient eligibility.

ELECTRONIC QUERIES

A natural extension of the logical checking system is the generation of electronic queries to the participating institutions. Once it has been determined that an error is not due to a central data entry error, a query is generated and sent either on paper or electronically to the institution. The query may ask for clarification on missing data values, values out of range, values that fail logical checks, or data that appear inconsistent. To be sure that the institution can interpret the query and respond accurately, it should contain the following:

- Protocol identifier
- Patient identifier
- Name of institution/individual to whom query is directed
- Return address of Coordinating Center (electronic or mail address)
- Date query sent
- Form/data item in question
- Specific description of the question(s)/clarification(s) to be answered
- Instructions on how to send response and time frame for submission

Some errors may be detected during manual rather than electronic review of the data, so it is useful to have a mechanism for generating a query manually. Using a query-generating program, the Data Coordinator could enter the protocol and patient identifier, the form/data item in question, and the text of the query. A printed/electronic copy could then be sent to the institution. Table 10.1 gives an example of a query sent from the Coordinating Center to a participating institution.

The Coordinating Center will need to have a mechanism for recording that a query has been sent and that a response has been received. It must also be possible to regenerate a query if the participating institution requests a second copy. Copies of all queries sent and responses received should be maintained

Table 10.1. Sample Query Letter.
Query Letter To: **Clinical Research Associate**   **General Hospital** **Protocol: 1234**   **Patient ID: 54321** **Date Sent: 6/6/96**
The follow-up form for visit 3 (on 4/21/98) had the following missing values: *Result of chest X ray* *Patient's performance status* *Pill count for self-administered medication* *Please submit a revised form with this data to the Trial Coordinating Center within 1 week. Thank you.* sent by   Jane Doe Data Coordinator Trials Center 1 High Street Anytown

at the Coordinating Center either on paper or electronically so that there is a complete record of the interactions between the institution and the Coordinating Center.

OVERDUE DATA REQUESTS

In any clinical trial it is important to ensure the submission of complete and timely data. However, in a busy clinic environment where the staff are involved in many clinical trials, it is often difficult to keep track of data submission requirements. A list of overdue data can be useful to the clinic staff and can be sent at regular intervals by the Coordinating Center. To be able to generate such a list, the Coordinating Center must be able to keep track of incoming data forms (whether paper or electronic) and to compare what has been received with a list of what is due. This can be done by creating the equivalent of a forms calendar for a particular trial, listing each data form and the submission schedule for each. This calendar can be stored on line. Note that if a form is submitted more than once during the trial, each occurrence needs to be represented in the calendar.

For each patient, the trigger date for the calendar is usually the date of registration onto the study. From that date it is possible to calculate the date that each subsequent form is due and therefore to detect when specific forms are overdue. A data request program can send listings of overdue forms to each participating institution on a regular basis. The data request needs to identify the institution, the protocol, the patient, the form(s) that are being requested, and instructions for submission. A typical data request for multiple studies is represented in Table 10.2. The specifications for the data request can be modified depending on the trial requirements. For some trials it may be beneficial to send out reminders *before* data are due. For that kind of report the same

---

Table 10.2. Sample Data Request.

Request for Overdue Data
Institution: General Hospital
Date List Generated: 6/6/98

Note: The following data forms are now overdue and should be submitted to:
          Trials Coordinating Center
          One Main Street
          Anytown
If data have already been sent, please check with the Trials Center before resubmitting

Protocol	Patient ID	Form	Remarks
1234	12001	Demographic Form	
1234	12003	Pathology Report	At time of relapse
		Query	Sent 4/1/98
1456	14090	Follow-up Form	Second visit
		Follow-up Form	Third Visit

kind of approach can be taken, but the list generated is based on a slightly different time frame than the one for overdue forms.

In any program requesting overdue forms, allowances should be made for delays in mail delivery. This allowance should be built into the program logic, and a statement should be made on the request that data recently sent should not be resent. Also, if the request asks for data that the site submitted long enough ago, that should have arrived at the Coordinating Center, the site should call the Coordinating Center to discuss this and not just resend the data. There may have been an error at the Coordinating Center that would not be corrected by resubmission of the form, and it could then continue to appear on future requests for data—and cause a lot of frustration!

CALENDARS AND REMINDERS

An alternative approach to ensuring that data are submitted on schedule is to develop a system that prospectively reminds institutions when specific forms/samples are due. These calendars (which can be generated at the time a patient is entered) can list all the required forms with a projected due date. The program could run at the Coordinating Center or at each of the participating institutions. The program would be more complex if data submission requirements are triggered by protocol events as well as by time frames. In

---

Table 10.3. Sample Forms Submission Calendar.

Projected Forms Submission Schedule
Protocol: 1234
Patient ID: 54321

Date entered: 3/1/98

Based on date of entry on study, the following are projected due dates for the study forms:

Form	Due date
Demographic Form	3/15/98
Pathology Report	3/15/98
Follow-up Form	6/1/98
	9/1/98
End of Treatment Form	12/1/98

this case, unless there is rapid submission of data to the Coordinating Center when such an event occurs, a version run at the sites will be able to produce more accurate output than one at the Coordinating Center. Table 10.3 gives an example of a forms calendar with projected due dates for forms for a patient.

STANDARD REPORTS

In any trial there will be a need to generate reports of the data. While some will be one-time-only reports that require special programming, many reports are run more than once during the course of the trial. The kinds of standard reports that might be useful are:

Accrual reports

Treatment safety reports

Administrative reports on study status

Interim statistical reports

It is efficient to write the programs for generating these reports and to retain them for future use. While the programs may need to be modified from time to time, this will save having to reproduce the code every time a report is needed. To ensure that the reports are useful, it is helpful to ask participants in the trial what kind of reports and data they would like to see. Note that Phase III trials are normally kept blinded (i.e., the outcome results are not made public) until the trial data has reached the design end points described in the study statistical section. A Data Monitoring Committee usually oversees these studies and adjudicates when the results can be unblinded.

INTERFACE WITH STATISTICAL SOFTWARE

Some database management systems come with software to interface with the major statistical software packages. However, if such an interface is not provided, one needs to be written so that data can be retrieved from the database and put into the statistical software. Most statistical packages require that data be formatted in a particular way, and there may be coding conventions that need to be observed. (e.g., representation of dates, representation of missing values or unknowns). There may need to be recodes of certain data items to comply with these conventions. The programmer will therefore need to be familiar with the requirements of the statistical software before the interface can be written, and the program needs to be thoroughly and rigorously tested to ensure that accurate data values are input into the statistical package.

PERFORMANCE MONITORING

In a multicenter clinical trial participating institutions are expected to accrue a certain number of patients on to each study and to submit the required data in a timely fashion. It is effective to have a system to monitor performance to ensure that the required standards are being met. This is especially important when a Coordinating Center works with the same group of institutions to conduct multiple trials. Since the participating centers must get feedback on their performance, the Coordinating Center must see whether participants are meeting required standards. The data generated for performance monitoring can also, if relevant, be used for future funding decisions for the sites.

The type of data reported will depend on the requirements of the trial. In a multi-trial environment where accrual, protocol compliance, timeliness of data submission, and publications are important, a performance report could be formatted as shown in Table 10.4. This kind of format allows performance to be monitored over time and would show either improvement, stability or decline over a three year period. This is important when trying to assess trends. The data also allow the participating institution (and the Coordinating Center) to see clearly where improvement is needed. In the institution in the example, accrual decreased dramatically in Year 2 but increased again in Year

Table 10.4. Sample Performance Evaluation Report.			
Annual Performance Report			
Institution: General Hospital Date: 12/31/98			
	Year 1	*Year 2*	*Year 3*
Accrual	53	21	65
Eligibility Rate	79%	85%	95%
Evaluable Cases	88%	90%	95%
Forms on Schedule	92%	92%	94%
Trial Publications	0	1	2

3. If there were no extraneous circumstances (e.g., lack of available proto-cols), the drop in Year 2 would be cause for concern. If this trend had contin-ued in Year 3, it would probably be a good indication that the institution had lost interest in participation. The institution was also having problems with eligibility and evaluability rates but was doing well getting forms in on sched-ule. Therefore any remedial training programs for this institution should be oriented toward eligibility and protocol compliance. The information about publications in this example would be useful to give credit for scientific lead-ership in publishing results of the trials.

Many performance monitoring programs would be much more complex than this example and may include algorithms for assessing relative perfor-mance of all participating sites, or details on other types of activities within the particular clinical trials program. However, this example illustrates the basic type of information that could be included in such a report.

Another decision to be made when implementing a performance monitor-ing system is whether any penalties are to be applied when performance is below required standards. If there are to be penalties, it is strongly recom-mended that they be defined and agreed to prior to review of the actual data. This will lead to a more objective review of each participant than if ad hoc decisions about penalties are made at the time the data are being reviewed. Normally a committee of people would be responsible for the review and assessment of performance. Typical sanctions could be warning letters, sus-pension of further accrual until certain conditions are met, or reduction in available funds. All are likely to have an impact and to improve future per-formance.

ON-LINE DATA LOOKUP

It is useful to have special purpose on-line data lookup utilities for members of the trial team. Participating institutions may want to be able to look up their accrual on a trial to date, or information about the last patient entered. Data Coordinators doing quality control of the submitted data may find it useful to have a program that will display key data items on the screen. Statisticians may want to review total accrual to date so that they can monitor when a study is reaching either an interim or final accrual goal. These types of utilities can lead to increased efficiency by making a routine query of the database easy to make. The following is an example of a summary of key data items for Data Coordinator review in a trial where survival data, number of visits, and total amount of drug administered are being updated:

Example: Summary Representation of Key Data Items

Protocol: 1234	Patient ID: 54321	
Date of last Contact: 6/6/98	Status: Alive	
Number of Visits at Time of Last Contact: 4		
Drug Administered:	Visit 1	300 mg
	Visit 2	300 mg
	Visit 3	150 mg
	Visit 4	300 mg
Total Drug to Date:		1050 mg
Record Last Updated:	7/31/98	

If the Data Coordinator is checking previously submitted data prior to updating new data and always wants to be certain that the new data are consistent with data previously submitted, rather than doing a detailed query to different parts of the database, this kind of lookup could be programmed so that it can be displayed by entering a minimal amount of information. For example, if the lookup program is called "summ" (for summary), it can be run for this patient by typing

summ 1234 54321

that is, by entering the name of the program, the protocol number and the patient id number. This algorithm can save a lot of time on the part of the Data Coordinator whenever new data is processed for a patient on study. There will be many frequently used lookup algorithms for a trial that could be programmed in this way.

APPOINTMENT SCHEDULES

Most clinical trials require that patients be seen at specified time intervals for tests, treatment, or completion of questionnaires. To ensure that clinics (and patients) are aware of the required schedule, it is possible to develop a computer program to generate patient-specific schedules at the time that a patient is entered on to the study. The schedule may need modification if there are delays in treatment or if events happen to change schedules, but it is a good starting point; if a copy is given to the patient, it helps them to understand the trial requirements and ensure compliance. Table 10.5 shows an example of the kind of reminder that could be given to patients when they are entered on a trial.

INVENTORY SYSTEMS

In large trials there is often a need to keep track of things such as drug supply, manuals, forms packets, and data. Programs can be written to keep track of the inventory and record items and quantities available, information about when supplies are distributed and to whom they were sent. Such a program can simplify some of the administrative requirements of a large trial.

DRUG ORDERS

Computer programs can be written to allow participants to order study drug for patients. Such a program should require entry of sufficient data to identify and validate the protocol, patient, and institution. A check can be made against the database to ensure that the patient is still eligible to receive drug and that the timing of the order is appropriate based on prior supply (i.e., that it is not too soon after the previous order). Once the order is approved, the shipping labels can be generated and a drug inventory log updated to indicate

---

Table 10.5. Sample Forms Submission Calendar.

Patient Name: John Doe   **Protocol:** Assessment of Quality of Life

Date entered: 6/1/98

FACT Questionnaire should be completed on the following dates during a clinic visit:

9/1/98
12/1/98
6/1/99
12/1/99

On 12/1/98 and 12/1/99 blood tests will also be done.

the amount of drug shipped, the date sent, and the relevant patient identification data.

COMMUNICATION TOOLS

In a multicenter trial it is important that all participants are kept well informed about the progress of the trial and about any problems that arise. While this communication can be done by regular mail or by fax, it can also be done by electronic mail (E-mail) if all participants are equipped to read and send E-mail. This method of communication can help to speed the dissemination of information and to ensure that all participants and the Coordinating Center have open lines of communication. Specialty user groups can be set up so that messages can easily be sent to a specific group of people (e.g., all the CRAs involved in a trial). With the expanding use of the Internet, many more people have access to electronic mail systems, and this makes E-mail a cost-effective means of communication for a clinical trial.

ON-LINE INFORMATION

As mentioned above, it is important to ensure that all participants are kept informed about the progress of the trial. If the participants have access to the Coordinating Center computer system or to a World Wide Web page set up by the Coordinating Center, it is possible to have routine information such as current accrual and projected closure dates available on-line for the participants. This kind of information is useful for participants when they are about to work up a patient for entry on to the study, since they can check to be sure that the trial is still open to new entries. Data and information that is required for annual IRB submissions could also be made available on-line, eliminating the need for each institution to request the information separately from the Coordinating Center.

With the growth of capability introduced by the World Wide Web, copies of documents such as protocols and manuals can be stored on-line and printed out locally by participants any time that they want them. This leads to savings in time and cost of mailing from the Coordinating Center.

SUMMARY

It is clear that computers can be used effectively to help manage clinical trials. The extent to which they are used will depend on the size, complexity, and duration of the trial and on the available expertise for programming. In an environment where many trials are coordinated, it is advisable to design data-driven software applications that can be used across multiple studies. While this increases the initial development time, it eliminates the need for extensive programming when a new trial is activated. The tools that can be developed can benefit the Coordinating Center, the institution, and the patient. Use of the Internet allows ready access to information and programs without the expense of developing a specialized communications system.

CHAPTER 11

Follow-Up and Closeout Phase

Most of the information in the previous chapters has described the first two phases in the life of a clinical trial—the design and development phase and the patient accrual phase. The third phase of a trial is the follow-up phase, where no new patients are being entered, but data are still being collected. The length of this phase will depend on the trial design and the study end points, and it will vary from trial to trial. In trials where the treatment period is short and the end points are assessed at the end of treatment or soon thereafter, the length of the follow-up phase will be very short, and the trial will be ready for analysis very soon after the last patient has finished treatment. An example of this would be a dermatology trial where a topical medication is being tested for its effect on clearing skin problems. The treatment effects will be seen quickly, and therefore there will not be a need for long-term follow-up of the patients. For a trial where the end points cannot be measured until many years after treatment ends, the follow-up phase will be lengthy. An example of this is a cancer trial where survival is the primary end point. It may be many years before there are sufficient follow-up data to be able to assess the impact of treatment on survival. The approach to collection of follow-up data will vary somewhat between these two types of trials.

TRIALS WITH A SHORT FOLLOW-UP PHASE

For the trials with a relatively short time between end of treatment and assessment of primary end points, it is essential to have a data collection system that will capture data quickly and have all treatment effect data in the database as soon as possible after each patient finishes treatment. The data collection for the follow-up phase will be minimal, with final case report forms being submitted as soon as possible after treatment ends and the end points have been assessed according to the protocol. With this type of trial it is critical to ensure that all data on all patients are coming in on schedule throughout the life of the protocol. Data need to be processed and edit queries dealt with rapidly. If this is done, the database will be kept current with queries resolved on an ongoing basis. After the last patient finishes treatment, there will be very little delay before the final data cleanup and analysis can be done.

TRIALS WITH A LONG FOLLOW-UP PHASE

When a clinical trial involves following patients for a long period of time, either with continuing treatment or after treatment ends, the issues of data collection take on a different focus. While it is still important to collect data and resolve queries as quickly as possible, there is not quite the same urgency as with a short-term trial. It is still true that the quality of the data is likely to be better if case report forms are completed on schedule and queries are sent without delay, and this is the goal that should be set for any trial. However, as will be discussed later in the chapter, the statistician and Study Chair are likely to generate additional questions when they start their work on the analysis, so there are likely to be queries generated well after a patient has stopped treatment. No matter how well you plan, this is always likely to happen.

In these trials it is possible that the staff originally involved in the trial at the participating sites will have moved on to other responsibilities or other places before the follow-up period is complete. If there is staff turnover, it is important for the Coordinating Center to maintain a roster of individuals who will submit the required follow-up data and can be contacted to answer questions. There may be a need to provide "follow-up" training or documentation for individuals who are not familiar with the trial.

The other important issue for these trials with long-term follow-up is ensuring that patients continue to be seen at the participating institution and that they do not become lost-to-follow-up. The best "cure" for lost-to-follow-up is prevention, and there are strategies that can be used when a patient is entered on the trial to minimize the risk of losing a patient. There are also several things that can be tried in efforts to locate a patient who does become lost-to-follow-up.

Minimizing Risk of Lost-to-Follow-Up

Before a patient consents to enter on to a clinical trial, they should be made aware of the follow-up requirements for a protocol. If it is known that the patient is planning to move away from the area after they have finished treatment, for example, or that they live far away and are making special arrangements to come for treatment, they are unlikely to be good candidates for long-term follow-up and therefore should probably not be entered on to the protocol.

If certain patients do consent to the protocol, then getting them involved and making them familiar with their actual follow-up schedule will be important. If they know when treatment is due, what tests will be required when, and the frequency of follow-up visits after treatment, they are more likely to comply. Patient education is therefore the first step to take in trying to ensure continued contact with the patients.

The CRA for the trial should be sure that the investigators also understand the importance of continued follow-up, and if the patient is moving away to another geographic area, or is going on an extended vacation, the CRA and the physician can try to identify a physician who is participating in the same trial in that area and who can assume responsibility for follow-up. If another trial physician cannot be identified, the patient could be asked to give the name and address of the physician they plan to see. This physician can be contacted, told about the requirements of the trial, and asked to provide any follow-up data that are needed. It will usually still be the responsibility of the institution that entered the patient to get the data and submit the case report forms to the Coordinating Center unless the patient is recognized by the Coordinating Center as officially transferred to become the responsibility of another participating institution.

During the treatment phase of the trial, it is important for the CRA or nurse to get to know the patient and let them know that they will be contacted periodically for routine follow-up. Patients will find it much more acceptable to be contacted by someone they know and someone who was involved in their treatment phase. The concept that there is someone who cares how they are doing can be an important one. If a patient is reluctant to return for scheduled follow-up visits post treatment, it may be appropriate to have the physician discuss the importance of the visits with the patient and try to encourage return visits.

As mentioned already, when the patient gives consent to enter on to the trial, the issue of follow-up should be discussed. This is a good time to ask the patient to give you the names, addresses, and telephone numbers of two or three people who could be contacted for information about the patient if the patient cannot be readily located. It is best to avoid spouse and dependent children who live with the patient, since, if the patient moves, they are also likely to move. Try to get names of close friends or other family members. The patient should be asked to let these people know that permission has been given for you to contact them if necessary.

Dealing with Lost-to-Follow-Up Patients

In clinical trials with long-term follow-up requirements it is almost inevitable that some patients will be lost for one reason or another. It is therefore important to have some procedures that can be implemented to try to locate these individuals. The following describe some possible methods of trying to locate a patient:

- If the patient was associated with a particular health plan or family doctor at the time of entry on the clinical trial, the health plan administrators or doctor's office may have more current information about the patient's whereabouts. If they have the information but are not willing to release it, they may be willing to contact the patient on your behalf and ask the patient to contact you.
- If the patient has an in-patient or out-patient medical record at your hospital or an administrative record for billing/insurance purposes, then there may be updated information in that record.
- Checking the telephone directory for the patient's name may give you a new number and address if the patient has moved within the same geographic area.
- Other options will probably depend on the country in which the search is being made. There may be national records kept of deaths or registries for specific illnesses such as cancer. If lists of eligible voters are maintained and available, they could also be searched. Central offices that keep track of licenses (e.g., drivers) may be another possible source. In some countries it may not be legal to ask for this kind of information, but in others you may be able to track the patient using one of these methods.

Remember that the patient has the right to refuse to be followed. At any time during the trial if any patients ask you not to contact them anymore, their wishes must be respected. This is one of the basic principles of the clinical trials process.

COLLECTING FOLLOW-UP DATA

Follow-up requirements for a trial should be kept to the minimum amount of data needed to assess the end points, and as with the data collected during the treatment period, the Coordinating Center should generate reminders to the institutions when follow-up is due. The follow-up schedule in the protocol should be realistic and, as far as possible, reflect actual clinical practice. For example, if it is normal for a certain group of patients to be seen every six months off protocol, then, unless the trial has very unusual requirements, the protocol should not require follow-up data to be submitted every month, since this would be clinically inappropriate and burdensome for the patient. The protocol should clearly identify the tests to be done during these visits and also the forms to be submitted. There should always be a mechanism for the institutions to report any long-term side effects of treatment. This is very important information to capture.

PREPARING FOR ANALYSIS

During the life of the trial, there will be several analyses of the data in the database. The analyses usually fall into two categories:

1. Administrative progress reports of the trial.
2. Formal interim analysis as defined in the statistical section of the protocol.

The analysis for routine progress reports will be primarily summaries of accrual, eligibility, and safety data. For Phase II trials there may also be reporting of outcome data. Phase III trials are usually monitored by the Data Monitoring Committee for the trial (see Chapter 2), and the formal analysis prepared for the Monitoring Committee will depend on the design and status of the trial. When any of these analyses are done, the Statistician will usually generate a list of questions about the data retrieved from the database, and the Data Coordinator will work with the Statistician to find answers. Sometimes this will involve trying to get additional data from the institutions. This is especially true when the final analysis is being done and the data are being reviewed in more detail than ever before. For the routine progress reports and the Data Monitoring reports, the queries are more likely to be to resolve inconsistencies in the database or to clarify some of the data. Since the preparation and content of the reports are usually the responsibility of the Statistician and the Study Chair rather than the data management staff, this chapter covers the purpose of the reports and describes the role of the Data Coordinator and CRA in preparing the reports. Other publications can be reviewed for input on the content and format of these reports.[1]

Freezing the Database

When reports are being generated, it is customary to "freeze" the database at a point in time and to use that "frozen" data set for generating the reports. During the time that the data are being retrieved for the analysis/report generation, no new data are added to the database, although corrections can be made if errors are found during the process. With most large trials, a cutoff date is set ahead of time, and all data received in the Coordinating Center by that date are processed and entered into the database. Once entered, the database is frozen, and the statisticians begin their retrievals. Any errors found are corrected, and then the final retrieval is done. Usually the statistician will need to run retrieval programs multiple times, sometimes to add additional data items to the retrieval and sometimes to incorporate data corrections. Note that all corrections should be made to the master database and not made by the statistician to the version of the file that they are using for analysis. Unless this procedure is followed, there is a risk that corrections made directly to the analysis files will not be applied to the database and will cause problems again the next time that the data are retrieved.

Once the final retrieval is done, the database can be updated again, and the Statistician can continue to work with the data file that has been retrieved. Usually the time that a data set is kept frozen can be kept short and will not interrupt the normal data management routines.

Using a frozen data set makes the analysis process easier, since the database is not constantly changing while the Statistician works on the data. If data updates continue, then new inconsistencies could be introduced into the data, and all the checking routines would have to be re-run every time the Statistician retrieves data. Also, in running institution performance monitoring in conjunction with the analysis, all institutions can be monitored by the same criteria—namely all data received as of a particular date. For any trial the Coordinating Center will need to decide whether to notify the institutions of any cutoff dates for monitoring. Letting the participants know about the dates will mean that there is more likelihood that a high proportion of required data will be received by the deadline. However, it can also mean that the Coordinating Center will have to process a large volume of data that arrives on the actual cutoff date, and this could extend the time that the database is frozen while the data are being quality controlled and entered into the computer.

Progress Reports

Routine progress reports are usually generated at regular intervals during a trial, and they provide information for the sponsor of the trial and for participants. As well as accrual, eligibility, and safety data, they may contain some data indicating institution performance in submission of data and in protocol compliance. The information will vary from trial to trial depending on the sponsor's requirements and the design of the trial. These periodic reports usually are in the same format each time, and software can be written to generate the report when required. The Data Coordinator as well as the statistician can review the report and check that there are no obvious inconsistencies. Usually the generation of these reports does not interrupt routine data processing at the Coordinating Center.

Formal Interim Reports

The reports reviewed by the Data Monitoring Committee for Phase III trials will contain much more information than the routine Progress Reports, and each report may focus on different types of information depending on the status of the trial. As well as reviewing data at the time of formal interim monitoring as defined in the trial, the committee may also review other aspects of the trial such as follow:

- Ongoing safety data for all patients to be sure that there are not unexpectedly high rates of serious or unusual Adverse Events on any of the treatment arms. The committee can recommend that the trial be suspended, terminated, or modified if the Adverse Event rate appears to be unacceptable.
- Problems with accrual that is slower than anticipated. The Data Monitoring Committee may make recommendations for ways to increase accrual or could recommend that a trial be terminated if the accrual rate does not increase after several attempts.

Usually the majority or all members of a Data Monitoring Committee are unconnected with the trial, and therefore any report that they review should probably contain background information about the objectives of the trial, previous decisions of the committee, the current status.

Final Analysis and Manuscript Preparation

Once the trial is ready for its final analysis, the final database cleanup will be done. It will be the Data Coordinator's responsibility to ensure that all missing data are requested from the participating sites and also that all the data from the Reference Centers are received and entered into the database. All clinical reviews of the data should be complete. The majority of trials are closed once their accrual goal has been reached and the data are mature. However, if a trial is closed early because of unexpected events, there will be increased urgency to finish these tasks so that the analysis can be done.

Once the database is as complete as possible, the statistician can begin the final analysis. The report generated from this analysis will usually be used to prepare one or more publications on the results of the trial, and often the statistician and the clinicians involved in preparing the analysis will generate lists of questions that need to be resolved, either from review of the case report forms that are in the Coordinating Center or by contacting the institutions with queries. It is unlikely that any new data could be consistently retrieved on all patients at this point of the trial, so it is impractical to think of collecting new data items that may have been overlooked when the original case report forms were designed! However, there may be some issues of clarification that require the Data Coordinator to contact the CRA at the institutions.

CLOSEOUT OF TRIAL

Many trials involve a formal "closeout" phase when the trial formally ends. There will be various responsibilities at the Coordinating Center and the participating sites during this phase.

Participating Sites

The sites will have to do some or all of the following:

- Submit all overdue case report forms and answers to queries.
- Return study materials such as drugs, devices, and shipping packages.
- File all trial records and arrange for storage as required by the sponsors or by federal regulations.

Coordinating Center

The Coordinating Center will normally be funded at a minimal level for a longer period than the field sites. During this time they will be responsible for mainly these items:

- Completion of all required analyses.
- Provision of all required materials to sponsor. This will usually include the original case report forms, all regulatory documents, an electronic

copy of the database, any inventory logs related to the trial, details of the randomization process, and any materials that were kept at the Coordinating Center, including unused drugs and other materials. The sponsor will also want a complete file of the protocol document and all amendments. Copies of all documentation and procedures manuals will also be required.

- Prepare a copy of the database for archiving. This will include a copy of the electronic files, related software, analysis files, and supporting documentation.

SUMMARY

While usually not as intense as the other phases of a trial, the follow-up and closeout phases are very important, particularly when the primary end points of the trial rely on having long-term follow-up data available for analysis. Much of the activity during this time is at the Coordinating Center, but the participating sites play an important role in ensuring that patients are followed and that any queries are answered. The Coordinating Center will be responsible for all analyses including the major analysis of the primary end points of the trial. The Center will usually also be involved in writing manuscripts for publication. Closeout responsibilities will depend on the sponsor or national requirements. It is important that the data and supporting documentation be archived safely so that the data from the trial can be accessed again in the future should the need arise.

REFERENCE

1. McBride R, Singer SW. Interim Reports, Participant Closeout, and Study Archives, *Controlled Clin Trials*, **16**, 137S–167S, 1995. (This article contains references to other relevant publications.)

CHAPTER 12

Training and Education

This chapter covers issues related to hiring, training, and continuing education for data management staff at the Coordinating Center as well as at the participating sites. Generic job descriptions for CRAs and Data Coordinators are included.

QUALIFICATIONS

Not everyone is suited to a career in data management, and because a lot of time and effort is invested in training new staff, it is important to try to get a good fit at the time of hiring. It can take several months to train a CRA or Data Coordinator to the point where this individual can work independently. The most important prerequisites for a CRA or a Data Coordinator are organizational skills and the ability to pay attention to detail. The trainee must be able to absorb sufficient knowledge about the trial data to be able to intelligently complete or process the forms. Increasingly as trials are becoming more computerized, it is also important that these individuals be able to use the computer systems to the extent required for the job. Given the nature of data management and the fact that there are frequent deadlines and urgent requests, it is further advisable that they be able to work under pressure.

In most clinical trials it is beneficial (and sometimes essential) that the data managers understand the clinical issues surrounding the trial and be able to interpret data in the medical records. A scientific background is therefore useful. Examples include individuals trained in nursing or medical records, or individuals who have completed college courses in biology or other sciences. Verbal and written communication skills are also important, since CRAs and Data Coordinators will have to communicate with other participants in the trial. The job descriptions for both CRAs and Data Coordinators will vary depending on the trial(s), the qualifications of the individual, and the systems in place at a specific location, but the following sections give some information about responsibilities that might be part of the job description of both positions.

JOB DESCRIPTION—CLINICAL RESEARCH ASSOCIATE

The primary responsibility of the CRA at the participating institution is to assist the investigator in the conduct of the trial at the participating site. It is important to note that the investigator is ultimately responsible for all trial-related activities at that site, although in reality many of them are completed by the CRA. This does not absolve the investigator from responsibility in ensuring that all federal and trial requirements are being met. The following are activities that could be the responsibility of a CRA depending on the situation at the local site and the qualifications of the CRA:

- Abstraction of data from medical records and completion of case report forms.
- Initial quality control of forms prior to submission.
- Submission of case report forms to the Coordinating Center in a timely way.
- Collection and submission of other trial related materials.
- Direct entry of data into computer system if remote data entry or a local computer system is being used.
- Preparation of protocols for initial submission to the local Ethics Committee or IRB.
- Submission of all protocol addenda/ revisions to the Ethics Committee.
- Submission of protocols for periodic re-approval by the Ethics Committee.
- Identifying possible trial patients.
- Verification of eligibility.
- Verification that patient has given consent to enter the trial.
- Randomization/registration of patients on to a trial.
- Interviewing patients.
- Maintaining protocol files, including complete copies of the protocol (and all revisions).
- Maintaining copies of all case report forms.

- Maintaining scheduling system for forms submission.
- Responding to requests and queries from the Coordinating Center.
- Preparing for audits.
- Communication with associated sites/departments where parts of the trial treatment is given.
- Preparation of materials for sending to Reference Centers.
- Training new staff.
- Preparing tools for clinicians to use during an appointment with a protocol patient to be sure that required tests are done and necessary data collected.
- Maintaining current versions of protocol in appropriate locations.
- Assisting with scheduling of tests and visits.
- Ensuring that appropriate drug logs are maintained when necessary.
- Monitoring inventory of any trial supplies (e.g., trial drugs) and reordering when needed.

Clearly not every CRA will have all these responsibilities, but they are all activities that need to be covered during the course of a clinical trial. If nurses are involved in the data management process, they are likely to also have additional responsibilities involving patient care and recording of data. The skills required for a CRA will depend in part on the responsibilities that are defined as part of the job description.

JOB DESCRIPTION—DATA COORDINATOR

The Data Coordinator in the Coordinating Center has a different, although related set of responsibilities. The Data Coordinator will not have any patient contact and does not have access to original medical records, except during visits to the participating sites. The following are responsibilities that could be part of the job description for a Data Coordinator:

- Quality control of submitted data, checking for completeness, and consistency.
- Generating queries to participating sites when necessary.
- Monitoring timeliness of data submission.
- Preparing data for entry into the computer system.

- Data entry.
- Definition of automated logical checks for a trial.
- Randomization/registration of patients.
- Monitoring regulatory compliance at participating sites.
- Maintaining protocol patient records at the Coordinating Center.
- Monitoring toxicities and Adverse Events.
- Preparing documentation for both the participating sites and the Coordinating Center.
- Conducting training workshops for participants.
- Assisting in the development of new forms.
- Assisting in defining the trial database.
- Participating in the source verification process either by conducting audits on site or by reviewing results of audits and ensuring that appropriate database corrections are done.
- Assisting statisticians with analyses.
- Generation of routine reports.
- Processing drug/supply orders from participating sites.
- Maintaining inventories of trial supplies at the Coordinating Center.

Again, usually not all these tasks will be done by one individual, but they are ones that need to be covered during the conduct of the trial. The most important function of the Data Coordinator is the quality control of submitted data, and this requires the ability to pay attention to detail and be accurate and consistent throughout the working day.

TRAINING AND ONGOING EDUCATION

Participants in a trial need to be trained so that they understand the trial requirements and how to meet them. Usually the sponsor or the Coordinating Center is responsible for developing training programs for the participants at the sites. There is a need both for initial training when the sites begin to enter patients on a trial and for continuing education about changes and refinements in the process. As new staff are added at sites, they will need to be trained. The following sections describe some training techniques that can be used.

WORKSHOPS

One way to ensure that all participants are fully informed about the requirements for the trial is to hold training workshops for either the CRAs or for all trial participants. The agenda for the workshop could include training on regulatory procedures, completion of forms, registration procedures use of trial software, and submission of materials to Reference Centers. This type of agenda is more relevant for CRAs than for the clinicians who enter and treat patients on the trial. However, if there are novel clinical procedures involved in the trial, it is also worthwhile holding sessions for the clinical staff. These sessions allow for questions from the participants as well as for structured presentations. The workshops can be held at the beginning of the trial and then periodically during the course of the trial to keep participants up to date in procedures. If necessary, they can be organized around specific areas that are causing problems.

VIDEOS

Another possible way of training the participants is to develop videos that cover training for specific aspects of the trial similar to the workshop agenda. While this can be more cost-effective because copies of the video can be mailed to the participants instead of requiring them all to travel to a workshop, it does not allow for interactive discussions between the trainers and those being trained. If videos are used, it is recommended that there be a user support telephone number that trainees can call with questions about the videos. Videos can become outdated quickly as trial procedures change, so it is important to be sure that only "current" instructions are included in any video used for training.

ON-SITE TRAINING

Training programs for participants can also be conducted on site. If a new CRA is hired at a site where there are already other CRAs experienced with the trial, then the experienced staff can be responsible for training the new CRA. However, if either the site is completely new to the trial or there are no experienced staff already at the site, the Coordinating Center could arrange some intense on-site training or the new CRA could be trained in part by CRAs at other sites, preferably those who are nearby geographically. A new

CRA can benefit from visiting a site where procedures are already in place for the trial or from having an experienced CRA from another site visit and explain how to set up procedures. It is essential that the clinical investigator(s) involved with the trial give support and training to the new CRA.

On-site training is also appropriate when a site is showing that it is having difficulties with particular aspects of the trial. A workshop can be targeted toward the problem areas in an attempt to overcome them.

MANUALS OF PROCEDURES

It is important to maintain consistency in process throughout the course of a trial, especially during a long trial where there may be changes in staff members at the sites or at the Coordinating Center. Manuals documenting policy and procedures for the trial need to be developed and circulated to participants. The manuals describe Standard Operating Procedures (SOPs) for the trial. The first versions of the manuals should be prepared prior to activation of the trial, and they should be updated as necessary during the course of the trial.

Manual for Participating Sites

The manual for the participating sites should provide information on all aspects of the trial and allow the participants to use it as a reference document to help them to fulfill their responsibilities during the trial and as a training manual for new staff. The following sections would normally be included in such a manual (or set of manuals):

> *Introduction.* This section would provide some background on the trial, the sponsor, the participants, and the Coordinating Center. Because the scientific rationale for the trial is included in the protocol, the manual would be more administrative in content than scientific.

> *Roster of participants.* There should be a section that has names, mailing addresses, telephone, fax, and, if relevant, E-mail addresses for all participants, including the sponsor, coordinating Center, Reference Centers, and anyone else associated with the trial. Usually at each participating site there is a Principal Investigator who is primarily responsible for the trial at that location. This person could be identified as well as the senior CRA. Names of contacts at the sponsoring organization and any Reference Cen-

ters should also be included, since this will simplify communications during the trial.

Regulatory requirements. The regulatory requirements for the trial should be clearly described and copies of any necessary forms included in the manual, along with instructions for how and when to complete them. If regulatory documents need to be submitted to the Coordinating Center, the schedule for submission should also be included.

Registration procedures. There should be detailed information on the registration process to supplement any information that is in the protocol document. The information should include hours of operation if registration relies on involvement of the coordinating staff. If an eligibility checklist is being used, it should be included along with instructions on completion. There should be information about any other data that will be required at the time of registration, such as demographic or stratification data. The rules applying to retrospective registrations should also be included— either that they are never allowed or a description of the circumstances under which they will be allowed. If the process is computerized in any way, there should also be a description of procedures to follow if the computer system is down.

Ordering of medications. If drugs or other medications/devices need to be ordered for a patient entered on the trial, instructions should be included about how to order them. There should also be information about disposal of any unused supplies at the end of the trial and about requirements for maintaining inventories of trial supplies. If the medication needs special preparation before use, this information should be included.

Data collection forms. All the required forms (or screens) for a study should be included in the manual along with detailed descriptions on how they should be completed. There should be clarification on the meaning of data items that can be confusing—such as if the trial is for patients who have recurrence of a disease and the forms ask for "Date of Initial Diagnosis," it should be explained whether this is date of first instance of the disease or date of the diagnosis that this recurrence was first noted. As the trial proceeds and the Coordinating Center processes submitted data, other clarifications will become apparent, and the Coordinating Center should update the participants' manual accordingly.

Data submission instructions. This section should explain how and when to submit all the required data.

Distributed data entry. If the data entry is being done at the sites, there will need to be detailed documentation on the installation, maintenance, and use of the software. This would normally be in a separate manual since a substantial amount of information is usually included.

Submission of materials. If tissue, blood, X rays, scans, or other materials need to be submitted to a Reference Center, there should be information on exactly what is needed, how it should be packaged and shipped, how shipping should be paid, and where and when it should be sent. If there are restrictions on when materials can be received at the Reference Center, then they should be clearly stated. For example, no shipments on Friday for Saturday delivery. There could also be information on whether or not the materials will be returned to the participating institution and how the results of the central review will be made known to the participants.

Data requests and queries. The manual should have information on how queries and requests will be sent by the Coordinating Center and how and when the sites should respond. Examples of the format of any queries and requests should be included along with an explanation of the content.

Adverse Event reporting. There should be definitions of the type of events that are considered serious and require rapid reporting. The instructions for reporting should be included, along with any form that needs to be completed and instructions on how to complete the form. Often these reports have to go to multiple places including the Coordinating Center, the sponsor, and the local Ethics Committee. Telephone/fax numbers and mailing addresses should all be included.

Performance monitoring. Any system for monitoring participation performance should be described. There should be a clear description of the expectations for each site and any penalties that will be applied if expectations are not met. The timing and format of the evaluation could also be discussed.

Progress reports. If the sites are to submit periodic progress reports on the trial at their institution, the format and schedule for submission should be discussed.

Communication. There should be information on contact people for various aspects of the trial along with details of how to contact them. In particular, there should be information on whom contact if there is a question about clinical care of a patient. Often more than one is person listed in case the primary contact cannot be reached in an emergency situation.

Preparation for an audit. There should be information about procedures for source verification of data, along with instructions on how the institution should prepare for such a visit. This would include details of what materials need to be made available and the format of the review. If a report is sent to the site after an audit, there should be information on whether a response is required and how to submit any revised data.

Maintenance of files. There should be details about what records need to be maintained and how long they have to be kept after the closure of the trial.

Fiscal Issues. During a trial the participating sites may be reimbursed for some or all of the trial related expenses. The manual should have information on the reimbursement rules and procedures.

Manual for Coordinating Center

The manual of procedures at the Coordinating Center will be more technical in content than the one for the participating sites. It will include details of the quality control process, the computer software being used, and procedures for analysis. The following are examples of sections that could be included:

Regulatory compliance. This section will describe the documents that need to be collected from each site, the time frame for collection, and the system for filing the information, whether electronic or paper based.

Registration/randomization procedures. The Coordinating Center manual will have details on how to process registrations and how to assign treatments. The documentation will also include information on the files that need to be maintained to document the process and about the confirmation of registration to be sent to the participating sites.

Quality control procedures. The Coordinating Center should develop an in-depth description of the procedures to be followed when reviewing the data submitted from the sites. This would include information about the checks to be done, procedures to be followed for generating queries, and conventions for when data should be returned unprocessed (e.g., missing values for key fields or wrong patient ID information on a form). There would also be descriptions of the process and order of review, the rules for assessing eligibility, protocol compliance, and outcome. There may also need to be explanation of the meaning of some of the data items and how they should be coded. Some of these procedures will be done manually, and some will be automated. If checks are automated, there will be rules for processing the error list that is generated by those checks. This part of the documentation needs to be extremely detailed and will probably be updated frequently in the early days of the trial as the initial data quality control is done.

Coding conventions. Part of the quality control documentation will include instructions on coding conventions for things like unknown values, tests not done, at missing data. For example, there may be some dates (e.g., date entered on study) where the exact date must be entered completely, and other dates (e.g., date of birth) where, if the day (or even month) is unknown, that it is acceptable and perhaps 15 is automatically entered for a missing day and 6 for a missing month. It is important to document these conventions at the start of the trial so that all data are processed in the same way. If certain fields come in with "other" coded as a value, it may be necessary to develop internal codes to identify "other" values that occur frequently (and that might be added to the form if it is to be revised at any time).

Data update procedures. This section would contain instructions on how to update data into the database. This would include documentation on how to run programs, who has the right to run the programs, and how to check the transactions to be sure they are correctly applied.

Trial procedures. There would also be documentation of other activities associated with the trial such as generating requests for missing data, communication with the reference centers, reporting of results, performance monitoring, and any other routine functions that are the responsibility of the Coordinating Center.

TRAINING COORDINATING CENTER STAFF

The training programs described so far have been aimed at the participants from the local sites. Most training of this type is conducted by the Coordinating Center staff. However, the staff in the Coordinating Center also need to be trained, and usually this training has to be done in-house. If a Coordinating Center handles many trials on an ongoing basis or has previously been responsible for a trial and has many procedures in place, experienced staff are usually responsible for training new staff. Depending on the complexities of the trial, it can take up to six months to fully train a Data Coordinator. It is important that all work be thoroughly checked until the trainer is confident that the new Data Coordinator is capable of working independently.

DATA MANAGEMENT COURSES

Very few schools/colleges/universities offer courses in data management, although there are some and it is worth checking in your area to see what is available. Biostatistics departments sometimes offer a course, and there are some programs for Clinical Research Associates. There are also short courses offered where there is intense training for a short period of time, usually up to a week. The absence of adequate general training courses in data management means that most of the training for new data management staff will be trial specific and run by the trial organizers. Check with your Coordinating Center or trial organizer to see if any courses might be available.

READING MATERIALS

Unfortunately, the literature on data management is also fairly sparse. Very few texts have been written on data management, although there are references to data management in some statistical and medical text books. There are also some journals that include articles on data management related issues. At the end of this book is a bibliography that lists articles and texts on data management. Some of them are fairly recent, and others were written several years ago but still have aspects that are relevant. Most of them are specific to one area of data management such as forms design, quality control, or data entry and are well worth reading if you are developing procedures for a trial.

It is also worthwhile to read about the experiences of others involved in clinical trials. The chances are that they have experienced the same or similar issues that are facing you, and it is beneficial to learn from their successes and mistakes and not to invest resources in reinventing procedures that have already been well thought out. It is hoped that this text will help you with the organization and conduct of your clinical trial.

SUMMARY

CRAs and Data Coordinators tend to come from diverse backgrounds. A medical or scientific background is an advantage, but the most important attributes for these positions are organizational and communication skills and the ability to pay attention to detail. Job descriptions can be extensive and varied and will depend on the local resources and requirements. Training and ongoing education are essential for any clinical trial, and the Coordinating Center should be responsible for organizing training programs and materials for trial participants. Most education is trial specific, but there are some courses and programs that cover aspects of data management as a career.

Bibliography

Adang RP, Vismans FJFE, Amergen AW, Talmon JL, Hasman A, Felndrig JA. Evaluation of Computerized Questionnaires Designed for Patients Referred for Gastrointestinal Endoscopy, *Int J Biomed Comput* **29**, 31–44, 1991.

Annerchiarico R, Moss AJ. The University Coordinated Drug Trial: An Organizational Schematic, *Controlled Clin Trials* **7**, 255, 1986.

Bagniewska A, Black D, Molvig K, Fox C, Ireland C, Smith J, Hulley S for the SHEP Research Group. Data Quality in a Distributed Data Processing System: The SHEP Pilot Study, *Controlled Clin Trials* **7**, 27–37, 1986.

Bailey LR. Distributed Data Entry, *1989 USIR Conference Proceedings*, Montreal 109–121, 1989.

Barker KN. Data Collection Techniques: Observation, *Am J Pharm*, **37**, 1235–1243, 1980.

Barnard PJ, Wright P, Wilcox P. Effects of Response Instructions and Question Style on the Ease of Completing Forms, *J Occup Psychol* **52**, 209–226, 1979.

Barwick JA, Foulkes MA, Elsinger JE, Rush RL. Addition of an Error Correction Facility to a Functioning Data Management System, *Controlled Clin Trials* **11**, 180–186, 1990.

Bjorn-Benson WM, Stibolt TB, Manske KA, Zavela KJ, Youtsey DJ, Buist AS. Monitoring Recruitment Effectiveness and Cost in a Clinical Trial, *Controlled Clin Trials* **14S**, 52–67, 1993.

Blackhurst DW, Maguire MG. Macular Photocoagulation Study Group. Reproducability of Refraction and Visual Acuity Measurements under a Standard Protocol, *Retina* **9**, 163–169, 1989.

Blessing JA, Bundy BN, Reese PA, Priore RL. Experience with the use of Generalized Database Management Systems in Cooperative Group Clinical Trials (A Project of the Gynecologic Oncology Group), *Controlled Clin Trials* **8**, 60–66, 1987.

Blumenstein BA. Verifying Keyed Medical Research Data, *Stat Med* **12**, 1535–1542, 1993.

Blumenstein BA, James KE, Lind BK, Mitchell HE. Functions and Organization of Coordinating Centers for Multicenter Studies, *Controlled Clin Trials* **16**, 4S–29S, 1995.

Blumenstein BA. The Relational Data Model and Multecenter Clinical Trials, *Controlled Clin Trials* **10**(4), 386–406, 1989.

Brant JD, Chalk SM. The Use of Automatic Editing in the 1981 Census, *J Roy Stat Soc (A),* **148(pt 2)**, 126–146, 1985.

Bulpitt CJ. *Randomized Controlled Clinical Trials*, Martinus Nijhoff Publishers, The Hague, 1983.

Buyse ME, Staquet MJ, Sylvester RJ. *Cancer Clinical Trials: Methods and Practice*, Oxford University Press, New York, 1984.

Byass P, Hanlon PW, Hanlon LCS, Marsh VM, Greenwood BM. Microcomputer Management of a Vaccine Trial, *Comp Biol Med,* **18(3)**, 179–193, 1988.

Cancer Research Campaign Working Party: Trials and Tribulations: Thoughts on the Organization of Multicenter Clinical Studies, *Brit Med J* **281**, 918–920, 1980.

Canner PL. Monitoring of the Data for Evidence of Adverse or Beneficial Treatment Effects, *Controlled Clin Trials* **4**, 467–483, 1983.

Chan LS, Portnoy B. Evaluation of Statistical Packages for Suitability for Use by Clinical Investigators in Medicine, *Comp Meth Prog Biomed,* **27**, 83–94, 1988.

Cheung S, Entine SM, Klotz JH. Microcomputer Voice-Response Telephone Entry for Balanced Clinical Trial Randomization, *J Med Sys*, **1**, 165–169, 1977.

Christiansen DH, Hosking JD, Dannenberg AN, Williams OD. Computer-Assisted Data Collection in Epidemiologic Research, *Controlled Clin Trials*, **11**, 101–115, 1990.

Connor PB, Newhouse MM, Grubb SC. Tracking, Accessing and Reporting on Edit Queries [abstract], *Controlled Clin Trials,* **11**, 272, 1990.

Crombie IK, Irving JM. An Investigation of Data Entry Methods with a Personal Computer, *Comp Biomed Res*, **19**, 543–550, 1986.

Cutter GR, Blanton MM, Perkins L. Distributed Data Analysis in Large Scale Trials: Should You or Shouldn't You? Proceedings of the Annual Meeting of the American Statistical Association, *Statistical Computing Section*, Las Vegas, August 5–8, 78–82, 1985.

Dassen WRM, Mulleneers R, Frank HLL. The Value of an Expert System in Performing Clinical Drug Trial, *Comput Biol Med* **21(4)**, 193–198, 1991.

Davis BR, Slymen DJ, Cooper CJ, Mutchler L. A Distributed Data Processing System in a Multicenter Clinical Trial, *Proceedings of the Annual Meeting of the American Statistical Association*, Statistical Computing Section, Las Vegas, August 5–8: 89–96, 1985.

de Dombal FT. Ethical Considerations Concerning Computers in Medicine in the 1980s, *J Med Ethics,* **13**, 179–184, 1987

DaBoer G, Bruce WR. Broadening Clinical Experience with Computer-Assisted Review, *MEDINFO*, **77**, 371–375, 1977.

DeMets DL, Mainert CL. Data Integrity, *Controlled Clin Trials*, **12**, 727–730, 1991.

DeMets DL et al. The Data Safety Monitoring Board and Acquired Immune Deficiency Syndrome, *Controlled Clin Trials*, **16**, 408–421, 1995.

DePauw M. Forms: Design and Content. In: *Data Management and Clinical Trials*, Rotmensz N, Vantongelen K, Renard J, eds. Elsevier, New York, 1989.

DuChene AG, Hultgren DH, Neaton JD, Grambsch PV, Broste SK, Aus BM, Rasmussen WL. Forms Control and Error Detection Procedures Used at the Coordinating Center of the Multiple Risk Factor Intervention Trial (MRFIT), *Controlled Clin Trials*, **7S**, 34–45, 1986.

Ederer F. The Statistician's Role in Developing a Protocol for a Clinical Trial, *Am Stat*, **33(3)**, 116–199, 1979.

Edvardsson B. Effect of Reversal of Response Scales in a Questionnaire, *Percept Mot Skills*, **50**, 1125–1126, 1980.

Eklundh KS, Marmolin H, Hedin CE. Experimental Evaluation of Dialogue Types for Data Entry, *Int J Man-Machine Studies*, **2**, 651–661, 1985.

Ellenberg S, Geller N, Simon R, Yusef S, eds. *Proceedings of Practical Issues in Data Monitoring of Clinical Trials*, Bethesda, MD, January 27–28, 1992. *Stat Med*, **12**, 415–616, 1993.

Etling LS, Bodey GP. Is a Picture Worth a Thousand Medical Words? A Randomized Trial of Reporting Formats for Medical Research Data, *Meth Inform Med*, **30**, 145–150, 1991.

Fedorowicz J, Haseman WD. Data Base Design for Infectious Disease Control. *J Med Syst*, **4(2)**, 107–119, 1980.

Finkelstein SM, Budd JR, Ewing LB, Wielinski CL, Warwick WJ, Kujawa SJ. Data Quality Assurance for a Health Monitoring Program, *Meth Inform Med*, **24(4)**, 192–196, 1985.

Fleming TR, DeMets DL. Monitoring of Clinical Trials: Issues and Recommendations, *Controlled Clin Trials*, **14**, 183–197, 1993.

Friedman LM, Furberg CD, Demets DL. *Fundamentals of Clinical Trials*, 2d ed. PSG Publishing, Littleton, MA, 1985.

Galitz WO. *Handbook of Screen Format Design*, QED Information Sciences; Wellesley, MA., 1985.

Gassman JJ, Owen WW, Kuntz TE, Martin JP, Amoroso WP. Data Quality Assurance, Monitoring and Reporting, *Controlled Clin Trials*, **16**, 104S–136S, 1995.

Gibson D, Harvey A, Everett V, Parmar MKB on Behalf of the CHART Steering Committee. Is Double Data Entry Necessary? The Chart Trials, *Controlled Clin Trials*, **15**, 482–488, 1994.

Goldman B, Jones O. Use of an Edit Feedback System in Data Collection Quality Control, *Am J Pub Health*, **68(7)**, 671–673, 1978.

Goodman PJ, Crowley JJ, Benson C. Creation of a Semiannual Report for a Multicenter Co-operative Clinical Trials Group, *Stat Med*, **11**, 1367–1376, 1992.

Gordis L. Assuring the Quality of Questionnaire Data in Epidemiologic Research, *Am J Epidemiol*, **109(1)**, 21–24, 1979.

Green SJ, Fleming TR, O'Fallon JR. Policies for Study Monitoring and Interim Reporting of Results, *J Clin Oncol,* **5**, 1477–1484, 1987.

Haakenson C, Akiyama T, Hallstrom A, Sather MR, FASHP for the CAPS Investigators. Masking Drug Treatment in the Cardiac Arrythmia Pilot Study (CAPS), *Controlled Clin Trials,* **17**, 294–303, 1996.

Habig RL, Thomas P, Lippel K, Anderson D, Lachin J. Central Laboratory Quality Control in the National Cooperative Gallstone Study, *Controlled Clin Trials,* **4**, 101–123, 1983.

Hamer DA. *1990 Data Collection and Processing Design. Proceedings of the Annual Meeting of the American Statistical Association,* August 17–20: 453–456, 1987.

Hawkins BS, Singer SW. Design, Development and Implementation of a Data Processing System for Multiple Controlled Trials and Epidemiologic Studies, *Controlled Clin Trials,* **7**, 89–117, 1986.

Hawkins BS. Data Monitoring Committees for Multicenter Clinical Trials Sponsored by the National Institutes of Health. I. Roles and Membership of Data Monitoring Committees for Trials Sponsored by the National Eye Institute, *Controlled Clin Trials,* **12**, 424–437, 1991.

Hawkins BS. A Survey of Data Monitoring Committees in NIH-Sponsored Multicenter Clinical Trials, *Controlled Clin Trials,* **11**, 273, 1990.

Hawkins BS, Gannon C, Hosking JD, James KE, Markowitz JA, Mowery RL. Report from a Workshop: Archives for Data and Documents from Completed Clinical Trials, *Controlled Clin Trials,* **9**, 19–22, 1988.

Hawkins BS. Controlled Clinical Trials in the 1980's: A Bibliography, *Controlled Clin Trials,* **12**, 1–272, 1991.

Heising KJ, Comstock GW. Response Variation and Location of Questions with a Questionnaire, *Int J Epidemiol,* **5**, 125–130, 1976.

Helms RW, McCanless I. The Conflict between Relational Databases and the Hierarchical Structure of Clinical Trials Data, *Controlled Clin Trials,* **11**, 7–23, 1990.

Hilner JE, McDonald A, VanHorn L, Bragg C, Caan B, Slattery ML, Birch R, Smoak CG, Wittes J. Quality Control of Dietary Data Collection in the CARDIA Study, *Controlled Clin Trials,* **13**, 156–169, 1992.

Holland WW. *Data Handling in Epidemiology,* Oxford University Press, London, 1970.

Horbar JD, Leahy KA for the Investigators of the Vermont–Oxford Trials Network. An Assessment of Data Quality in the Vermont–Oxford Trials Network Database, *Controlled Clin Trials,* **16**, 51–61, 1995.

Hosking JD, Newhouse M, Bagniewska A, Hawkins BS. Data Collection and Transcription, *Controlled Clin Trials,* **16**, 66S–103S, 1995.

Hosking JD, Rochon J. A Comparison of Techniques for Detecting and Preventing Key-Field Errors, *Proceedings of the Annual Meeting of the American Statistical Association,* Statistical Computing Section, Cincinnati, August 16–19, 82–87, 1982.

Hosking JD, Rochon J, Scott P. Data Base Closure: The Transition from Data Processing to Data Archiving, *Proceedings of the Annual Meeting of the American Statistical Association*, Statistical Computing Section, Detroit, MI, August 10–13, 246–251, 1981.

Hulley SB, Cummings SR. *Designing Clinical Research: An Epidemiologic Approach*, William and Wilkins, Baltimore, MD, 1988.

Jefferys JL, Lee TC, Tonascia J. Distributed Processing System in Medical Clinical Trials. *Proceedings of the Annual Meeting of the American Statistical Association*, Statistical Computing Section, Las Vegas, August 5–8, 83–88, 1985.

Joseph M, Schoeffler K, Doi PA, Yefko H, Engle C, Nissman E. An Automated COSTART Coding Scheme, *Drug Inform J*, **25**, 97–108, 1991.

Karrison T. Data Editing in a Clinical Trial, *Controlled Clin Trials*, **2(1)**, 15–29, 1981.

Keltner JL, Johnson CA, Beck RW, Cleary PA, Spurr JO, Optic Neuritis Study Group. Quality Control Functions of the Visual Field Reading Center (VFRC) for the Optic Neuritis Treatment Trial (ONTT), *Controlled Clin Trials*, **14**, 143–159, 1993.

King C, Manire L, Strong R. *Comparing Data Management Systems in Clinical Research: A 1983 Survey*, Health Systems Project, Harvard School of Public Health, Boston, 1983.

Knatterud GL, Forman SA, Canner PL. Design of Data Forms, *Controlled Clin Trials*, **4**, 429–440, 1983.

Knatterud GL. Methods of Quality Control and of Continuous Audit Procedures for Controlled Clinical Trials, *Controlled Clin Trials*, **1**, 327–332, 1981.

Krischer JP, Hurley C, Pillamarri M, Pant S, Bleichfeld C, Opel M, Shuster JJ. An Automated Patient Registration and Treatment Randomization System for Multicenter Clinical Trials, *Controlled Clin Trials*, **12(3)**, 367–377, 1991.

Krol WF. Closing Down the Study, *Controlled Clin Trials*, **4**, 505:512, 1983.

Kronmal RA, Davis K, Fisher LD, Jones RA, Gillespie MJ. Data Management for a Large Collaborative Clinical Trial. (CASS: Coronary Artery Surgery Study), *Comp Biomed Res*, **11**, 553–566, 1978.

Lange N, MacIntyre J. A Computerized Patient Registration and Treatment Randomization System for Multi-institutional Clinical Trials, *Controlled Clin Trials*, **6**, 38–50, 1985.

Lee JY. An Interactive System for Clinical Trials, *Comp Applications in Med Care*, 404–405, 1979.

Levine RJ. *Ethics and Regulation in Clinical Research*, 2d ed, Urban & Schwarzenberg, Baltimore, MD, 1986.

LoPresti FF, Clarke EJ, Wilkins PC. Functional Comparisons between Distributed and Centralized Clinical Trial Data Processing Systems [abstract], *Controlled Clin Trials*, **9(3)**, 266, 1988.

Lundberg ED, McBride R, Rawson TE, Mauritsen R, Ormond TH, Fisher LD, Kronmal RA, Gillespie MJ. C2: A Data Base Management System Developed for the

Coronary Artery Surgery Study (CASS) and other clinical studies, *J Med Syst*, **6(5)**, 501–518, 1982.

Mangione TW, Hingson R, Barrett J. Collecting sensitive data: A Comparison of Three Survey Strategies, *Social Methods and Research*, **10(3)**, 337–346, 1982.

Marinez YN, McMahon A, Barnwell GM, Wigodsky HS. Ensuring Data Quality in Medical Research through an Integrated Data Management System, *Stat Med*, **3**, 101–111, 1984.

McBride R, Singer SW. Interim Reports, Participant Closeout, and Study Archives, *Controlled Clin Trials*, **16**, 137S–167S, 1995.

McDonald CJ. Protocol-Based Computer Reminders, the Quality of Care and the Non-perfectibility of Man, *N Eng J Med*, **295**, 1351–1355, 1976.

McFadden ET, LoPresti F, Bailey LR, Clarke E, Wilkins PC. Approaches to Data Management, *Controlled Clin Trials*, **16**, 30S–65S, 1995.

McShane LM, Turnbull BW. Optimal Checking Procedures for Monitoring Laboratory Analyses, *Stat Med*, **11**, 10, 1343–1357, 1992.

Megill D, Rowland S. Data Edit Error Analysis, *Proceedings of the Annual Meeting of the American Statistical Association*, August 17–20, 154–159, 1987.

Meinert CL. Clinical Trials and Data Integrity, *Controlled Clin Trials*, **1**, 189–192, 1980.

Meinert CL, Heinz EC, Forman SA. Role and Methods of the Coordinating Center, *Controlled Clin Trials*, **4**, 355–375, 1983.

Meinert CL. *Clinical Trials: Design, Conduct and Analysis*, Oxford University Press, New York, 1986.

Moher D, Jadad AR, Nichol G, Penman M, Tugwell P, Walsh S. Assessing the Quality of Randomized Controlled Trials: An Annotated Bibliography of Scales and Checklists, *Controlled Clin Trials*, **16**, 62–73, 1995.

Moye LA. Central Laboratory Sampling Plans and Quality Control in Clinical Trials, *Controlled Clin Trials*, **12**, 761–767, 1991.

Mullooly JP. The Effects of Data Entry Errors: An Analysis of Partial Verification, *Comp Biomed Res*, **23**, 259–267, 1990.

Multiple Risk Factor Intervention Trial. Quality Control of Technical Procedures and Data Acquisition, *Controlled Clin Trials*, **7s**, 1–202, 1986.

Neaton JD, Duchene AG, Svndsen KH, Wentworth D. An Examination of the Efficiency of Some Quality Assurance Methods Commonly Employed in Clinical Trials, *Stat Med*, **9**, 115–124, 1990.

Newhouse MM. Data Entry Design and Data Quality [abstract], *Controlled Clin Trials*, **6**, 229, 1985.

Nilan-Weiss J, Azen SP, Odom-Maryon T, Lui F, Hagerty C. A Microcomputer-Based Distributed Data Management System for a Large Cooperative Study of Transfusion Associated Acquired Immunodeficiency Syndrome, *Comp Biomed Res*, **20**, 225–243, 1987.

Overton HH. Perceptions of the Coordinating Center as Viewed by a Clinic Coordinator, *Controlled Clin Trials*, **1**, 133–136, 1980.

O'Fallon JR. Data Management and Quality Control. In, *Statistics in Medical Research*, Mike V, Stanley KE, eds. Wiley, New York, 1982.

Palit CD, Sharp H. Microcomputer-Assisted Telephone Interviewing, *Sociol Meth Res*, **12**, 119–142, 1983.

Payne SL. *The Art of Asking Questions*, Princeton University Press, Princeton, NJ, 1951.

Perkins LL, Cutter GR, Wagenknecht LE, Savage PJ, Dyer AR, Birch R. Distributed Data Analysis in a Multicenter Study: The CARDIA Study, *Controlled Clin Trials*, **13(1)**, 80–90, 1992.

Pocock SJ. *Clinical Trials: A Practical Approach*, J Wiley, New York, 1983.

Poe GS, Seeman I, McLaughlin J, Mehl E, Dietz M. Effects on Level and Quality of Response in the Inclusion of "Don't Know" Boxes in Factual Questions in a Mail Questionnaire, *Proceedings of the Annual Meeting of the American Statistical Association*, August 17–20, 277–281, 1987.

Pradhan EK, Katz J, LeClerq SC, West KP. Data Management for Large Community Trials in Nepal, *Controlled Clin Trials*, **15**, 220–234, 1994.

Prud'homme GJ, Canner PL, Cutler JA for the Hypertension Prevention Trial Research Group. Quality Assurance and Monitoring in the Hypertension Prevention Trial, *Controlled Clin Trials*, **10s**, 84–94, 1989.

Reddy NP, Kesavan SK. A Simple Coding Method for Computer Storage and Handling of Drug Information, *Int J Bio-Med Computing*, **18**, 131–134, 1986.

Renard J, Van Glabbeke M. Quality Control and Data Handling. In, *Data Management and Clinical Trials*, Rotmensz N, Vantongelen K, Renard J, eds. Elsevier, New York, 1989, pp 147–162.

Reynolds-Haertle RA, McBride R. Single vs Double Data Entry in CAST, *Controlled Clin Trials*, **13**, 487–494, 1992.

Rifkind BM. Perceptions of the Coordinating Center as Viewed by a Project Officer, *Controlled Clin Trials*, **1**, 137–141, 1980.

Rondel RK, Varley SA, Webb CF. *Clinical Data Management*, Chichester, Wiley, New York, 1993.

Rosendorf LL, Dafni V, Amato DA, Lunghofer B, et al. Performance Evaluation in Multicenter Clinical Trials: Development of a Model by the AIDS Clinical Trials Group, *Controlled Clin Trials*, **14**, 523–537, 1993.

Rotmensz N, Vantongelen K, Renard J, eds. *Data Management and Clinical Trials*, Elsevier, New York, 1989.

Royston P, Bercini D. Questionnaire Design Research in a Laboratory Setting: Results of Testing Cancer Risk Factor Questions, *Proceedings of the Annual Meeting of the American Statistical Association*, August 17–20, 829–833, 1987.

Sackett G. *Observing Behavior II. Data Collection and Analysis Methods*, University Park Press, Baltimore, MD, 1978.

Saltzman A. Adverse Reaction Terminology Standardization: A Report on Schering-Plough's Use of the WHO Dictionary and the Formation of the WHO Adverse

Reaction Terminology Users Group (WUG), *Consortium Drug Inform J*, **10**, 35–41, 1985.

Schwartz RP. Maintaining Integrity and Credibility in Industry-Sponsored Clinical Research, *Controlled Clin Trials*, **12**, 753–760, 1991.

Senn, S. *Cross-over Trials in Clinical Research*, J Wiley, New York, 1993.

Shapiro SH, Louis TA, eds. *Clinical Trials: Issues and Approaches*, Dekker, New York, 1983.

Siebert C, Clark CM. Operational and Policy Considerations of Data Monitoring in Clinical Trials: The Diabetes Control and Complication Trial Experience, *Controlled Clin Trials*, **14**, 30–43, 1993.

Sills J. World Health Organization Adverse Reaction Terminology Dictionary, *Drug Inform J*, **23**, 211–216, 1989.

Silverman WA. *Human Experimentation: A Guided Step into the Unknown*, Oxford University Press, New York, 1985.

Singer SW, Meinert CL. Format-Independent Data Collection Forms, *Controlled Clin Trials*, **16**, 363–376, 1995.

Spencer BD. Optimal Data Quality, *J Am Stat Assoc*, **80**, 564–573, 1985.

Spilker B. *Guide to Clinical Studies and Developing Protocols*, Raven Press, New York, 1984.

Spilker B, Schoenfelder J. *Data Collection Forms in Clinical Trials*, Raven Press, New York, 1991.

Spilker B, Cramer J. *Patient Recruitment in Clinical Trials*, Raven Press, New York, 1992.

Spilker B. *Guide to Planning and Managing Multiple Clinical Studies*, Raven Press, New York, 1987.

Starmer CF, Smith DAH, Wells JS, Wright BC. Problems in Data Management When Studying Chronic Illness, *J Med Syst*, **5(4)**, 271–280, 1981.

Stellman SD. The Case of the Missing Eights: An Object Lesson in Data Quality Assurance, *Am J Epidemiol*, **129(4)**, 857–860, 1989.

Sudman S, Bradbum NM: Asking Questions. *A Practical Guide to Questionnaire Design*, Jossey-Bass, San Francisco, 1983.

Sweet F. *What, If Anything, Is a Relational Database?* Datamation, 118–124, 1984.

Taylor DW, Bosch EG. The Datafax Project [abstract], *Controlled Clin Trials*, **11(4)**, 296, 1991.

Taylor W, Bosch EG. DataFax Evaluated [abstract], *Controlled Clin Trials*, **12**, 644, 1991.

van der Putten E, van der Velden JW, Siers A, Hamersma EAM. A Pilot Study on the Quality of Data Management in a Cancer Clinical Trial, *Controlled Clin Trials*, **8**, 96–100, 1987.

Verter J. How Much Data Should We Collect in a Randomized Clinical Trial, *Stat Med*, **9**, 103–113, 1990.

Voynick IM, Makuch RW. The Design and Implementation of a Software System for Clinical Studies: An Illustration Based on the Needs of a Comprehensive Cancer Center, *J Med Syst*, **12(5)**, 295–304, 1988.

Weiner JM. *Issues in the Design and Evaluation of Medical Trials*, GK Hall, Boston, 1979.

Weiss DG, Williford WO, Collins JF, Bingham SF. Planning Multicenter Clinical Trials: A Biostatistician's Perspective, *Controlled Clin Trials*, **4**, 53–64, 1983.

Wright P, Haybittle J. Design of Forms for Clinical Trials [3-part article], *Br Med J*, **2**, 529–530, 590–592, 650–651, 1979.

Index

WILEY SERIES IN PROBABILITY AND STATISTICS

Probability and Statistics
  ANDERSON · An Introduction to Multivariate Statistical Analysis, *Second Edition*
  *ANDERSON · The Statistical Analysis of Time Series
  ARNOLD, BALAKRISHNAN, and NAGARAJA · A First Course in Order Statistics
  BACCELLI, COHEN, OLSDER, and QUADRAT · Synchronization and Linearity:
    An Algebra for Discrete Event Systems
  BARTOSZYNSKI and NIEWIADOMSKA-BUGAJ · Probability and Statistical Inference
  BERNARDO and SMITH · Bayesian Statistical Concepts and Theory
  BHATTACHARYYA and JOHNSON · Statistical Concepts and Methods
  BILLINGSLEY · Convergence of Probability Measures
  BILLINGSLEY · Probability and Measure, *Second Edition*
  BOROVKOV · Asymptotic Methods in Queuing Theory
  BRANDT, FRANKEN, and LISEK · Stationary Stochastic Models
  CAINES · Linear Stochastic Systems
  CAIROLI and DALANG · Sequential Stochastic Optimization
  CHEN · Recursive Estimation and Control for Stochastic Systems
  CONSTANTINE · Combinatorial Theory and Statistical Design
  COOK and WEISBERG · An Introduction to Regression Graphics
  COVER and THOMAS · Elements of Information Theory
  CSÖRGŐ and HORVÁTH · Weighted Approximations in Probability Statistics
  *DOOB · Stochastic Processes
  DUDEWICZ and MISHRA · Modern Mathematical Statistics
  DUPUIS and ELLIS · A Weak Convergence Approach to the Theory of Large Deviations
  ETHIER and KURTZ · Markov Processes: Characterization and Convergence
  FELLER · An Introduction to Probability Theory and Its Applications, Volume 1,
    Third Edition, Revised; Volume II, *Second Edition*
  FREEMAN and SMITH · Aspects of Uncertainty: A Tribute to D. V. Lindley
  FULLER · Introduction to Statistical Time Series, *Second Edition*
  FULLER · Measurement Error Models
  GHOSH, MUKHOPADHYAY, and SEN · Sequential Estimation
  GIFI · Nonlinear Multivariate Analysis
  GUTTORP · Statistical Inference for Branching Processes
  HALD · A History of Probability and Statistics and Their Applications before 1750
  HALL · Introduction to the Theory of Coverage Processes
  HANNAN and DEISTLER · The Statistical Theory of Linear Systems
  HEDAYAT and SINHA · Design and Inference in Finite Population Sampling
  HOEL · Introduction to Mathematical Statistics, *Fifth Edition*
  HUBER · Robust Statistics
  IMAN and CONOVER · A Modern Approach to Statistics
  JOHNSON and KOTZ · Leading Personalities in Statistical Sciences: From the
    Seventeenth Century to the Present
  JUREK and MASON · Operator-Limit Distributions in Probability Theory

*Now available in a lower priced paperback edition in the Wiley Classics Library.

*Now available in a lower priced paperback edition in the Wiley Classics Library.

*Now available in a lower priced paperback edition in the Wiley Classics Library.

*Now available in a lower priced paperback edition in the Wiley Classics Library.

*Now available in a lower priced paperback edition in the Wiley Classics Library.